Recording Voiceover

Bound to Create

You are a creator.

Whatever your form of expression — photography, filmmaking, animation, games, audio, media communication, web design, or theatre — you simply want to create without limitation. Bound by nothing except your own creativity and determination.

Focal Press can help.

For over 75 years Focal has published books that support your creative goals. Our founder, Andor Kraszna-Krausz, established Focal in 1938 so you could have access to leading-edge expert knowledge, techniques, and tools that allow you to create without constraint. We strive to create exceptional, engaging, and practical content that helps you master your passion.

Focal Press and you.

Bound to create.

We'd love to hear how we've helped you create. Share your experience:

www.focalpress.com/boundtocreate

Recording Voiceover
The Spoken Word in Media

Tom Blakemore

Focal Press
Taylor & Francis Group

NEW YORK AND LONDON

First published 2015
by Focal Press
70 Blanchard Road, Suite 402, Burlington, MA 01803

and by Focal Press
2 Park Square, Milton Park, Abingdon, Oxon OX14 4RN

Focal Press is an imprint of the Taylor & Francis Group, an informa business

Library of Congress Cataloging in Publication Data
Blakemore, Tom.
   Recording voiceover: the spoken word in media/Tom Blakemore.
   pages cm
   Includes bibliographical references.
   1. Voice-overs—Vocational guidance. 2. Sound—Recordings and
   reproduction—Handbooks, manuals, etc. 3. Sound recordings—
   Production and direction—Handbooks, manuals, etc. I. Title.
   PN1990.9.A54B53 2015
   621.382′8—dc23
   2014039136

ISBN: 978-0-415-71608-6 (hbk)
ISBN: 978-0-415-71609-3 (pbk)
ISBN: 978-1-315-88017-4 (ebk)

Typeset in Times and Trade Gothic
by Florence Production Ltd, Stoodleigh, Devon, UK

Printed and bound in the United States of America by Sheridan Books, Inc. (a Sheridan Group Company).

For Jim Thomson:
recording engineer extraordinaire,
inventor, friend, and the best mentor
an aspiring young engineer could
possibly have had. This is for you—
thanks, Jim, we miss you.

Contents

Author's Biography

Tom Blakemore has been an active audio engineer for over thirty years, working in film, television, commercial, and corporate communications as a supervising sound editor and mixer. He has edited and mixed for virtually all of the major broadcast and cable outlets and is a veteran of thousands of national radio and television commercials. His film work includes Emmy Award-winning documentaries, Academy Award nominees, Directors Guild of America Best Documentary winners, and Audience Award winners at the Toronto, Chicago, and Amsterdam Film Festivals. Tom lives in Chicago, where he is an Adjunct Professor at Tribeca Flashpoint Media Arts Academy, teaching film sound, and works as a freelance audio engineer. He is a member of the Motion Picture Sound Editors and the Audio Engineering Society.

Foreword

If you have a strong interest in recording and would like to learn how those great-sounding voices that you hear on radio spots, television commercials, and even your favorite animated movies come together, then Tom Blakemore's *Recording Voiceover* is your pathway into the microphone, through the tape rollers, and over the record heads.

Tom has written the first book I have ever read on the subject of *recording* voiceover itself. Accurate and thorough, it is clear that only a professional sound engineer, with years of recording experience and a teaching background, could have written it. Tom gives his readers a clear explanation of the steps it takes to prepare for a voiceover session, from script preparation to picking the proper microphone.

In this book, Tom gives you insights on what is expected from a voiceover recording engineer. Will the recording session be for a radio spot, television commercial, or a film trailer? Choice of microphone and its placement will vary for each of these projects. Keep in mind, a talented engineer had to capture all the grit and gravel of the famous voiceover artist, Don La Fontaine, recording one of the most recognized lines in film trailer history: "*In a world gone mad.*"

You are entering into a very specialized area of the entertainment industry that operates behind the scenes, with no glamour and no paparazzi. Not even an article in *Mix Magazine* in sight! Your new clients have all gathered into a recording studio, where they have booked you to capture the smooth, velvety tones of their voiceover talent for a deodorant commercial. A director, a writer, and a producer are all hanging on to every word, listing intently to every phrase, every breath, and all that velvet is passing through your perfectly positioned, very expensive, German microphone, which itself is transferring that velvet through your very expensive Canare mic cable, into your very expensive analog microphone preamp and compressor, through your hands, into your DAW, and, *oh wow*, this talent is sounding like a god or goddess, with perfect diction! You play the take back, every syllable captured, no esses, no breath pops, no P pops. This is what you live for—a perfect recording. For that moment, you are a star. The director reaches for the talk back and gives you a perfectly deadpan: "That was perfect, let's go again!" Welcome back to reality, and your world of voiceover recording.

Maybe you are reading this book because you want information on what it takes to build your own recording space. Tom's book covers studio design and the proper materials required for construction of the studio. Pay special attention to the record room, as this is where all your great recordings will start, and, if you can, build it with a high ceiling. Make sure you have a control room that is big enough for your future clients. Here is an example of a session that you might someday experience.

What could be simpler then recording a regional bank spot or a tag line for an automobile commercial? These projects involve many people, and, come the day of the recording for these spots, you might just find those many people all descending into your studio! This might include the advertising representatives, producers of the commercial, and the copywriters of the spot. Now, bring in the director and voiceover talent, and you have yourself a full house. Moments later, a few executives from the company that these spots are for have arrived with their wives, entirely unannounced. You can see how quick your control room is filling up!

Tom gives such great advice in his book, but, to me, his most important is: Be prepared and ready to roll at all times. And, I would like to add, never be late for a session. I was in a session once where an actor had arrived early, went straight for the mic and started riffing on anything that came to his head, testing out a new stand-up routine on fresh ears. Some of those bits and pieces could actually be used in the movie he was there to record for, all because we were record-ready and punched in way before the talent's call time.

The studio I work in is capable of animation recording, VO recording, and ADR, but each of these configurations requires different record-room seating and microphone arrangements. We call it "turning the room around." On Monday, I may be recording an ADR session and, on Tuesday, a four-microphone setup for our animation clients. Wednesday will be back to ADR for a video game involving forty actors. Turning the room around for each of these sessions can take up to two hours. I always try to have the studio record-ready the evening before the actual session; yes, I might have to stay late! And that also includes making sure my Pro Tools session template is named properly and routed correctly in my I/O setup. I double check to make sure all the microphone signals are truly routed to the proper record channels (DAWs can, and will, boot up cranky).

Another reason I like to be ready early is so, when the client arrives, we can have some face time, talk about family and how the kids are, and then move on to how the show is coming along. Keep it relaxed! Moving chairs, go-betweens, and boom stands while trying to have a conversation is a bit like talking to a mechanic while he's under your car.

My favorite chapter in *Recording Voiceover* is Chapter 11: Recording for Games and Animation. I would like to offer an exercise that will forever keep you from getting in trouble with your talent in the future!

I had the pleasure of working with the late Robin Williams on several projects. He was fantastic, but notorious for being able to go from a whisper to a scream without notice. Robin would be listening to his voice through headphones while recording, and any microphone distortion would stop him right away, thus taking him out of his brilliant thought process and interrupting the magic. It never sat well with him.

If you are currently working in a studio and have not been subject to such a dynamic voice talent, then this exercise may help you. Put up a Neumann U87. If the studio does not own one, ask if they can rent one for you. Ask a friend with a big voice to join you at the studio and explain to him that this is a microphone distortion test. Your friend will have to whisper, then talk in his normal voice, then scream, then talk in his normal voice, then scream, then whisper and scream. Have him repeat this for as long as he can. This is the only way to learn how the mic and yourself will respond under these extreme level changes. This is also the time to dial in the proper levels for your microphone preamps and compressor settings. Document your final settings, so that, when the day comes, you can dial them in. Remember . . . no distortion!

I use the U87 in this test because I truly believe this will be your go-to microphone for many of your voiceover recording sessions.

Vince Caro, my friend and colleague from Pixar Animation Studios, writes a terrific section about how he plans and prepares for his sessions, from a large orchestra to a multiple-microphone animation session for Pixar. Vince explains his use of vintage microphones, and how he incorporates them into today's recording sessions.

When I had to say goodbye to our analog mixing console and enter into the world of Pro Tools and digital recording, Vince was there to help me through the transition with his invaluable knowledge of the digital recording medium.

I want to thank Tom Blakemore for offering me the opportunity to write this foreword for *Recording Voiceover*. Tom's love for recording and his love for teaching shine throughout his book.

Doc Kane
ADR and Dialogue Mixer, The Walt Disney
Studios, Burbank, California

Preface

In the course of my recording and teaching career, I've constantly been amazed at the lackadaisical attitude that many people have in recording a voiceover. This is equally true for students and professionals alike. Although they may obsess for hours over what microphone to use to record a particular guitar or vocalist, in many, many instances, when it comes to voiceover recording, they will set up the closest mic at hand and then sit back as the actor dutifully reads the copy. There is often little thought given to the peculiarities of the voiceover session, and, although these people may give a commendable level of attention to detail in a music session, for the voice session, it seems they are barely paying attention. As with anything else that we create, paying attention to the minutiae of the job at hand is the only way to arrive at a superior result.

When it comes to voiceover recording, this extends beyond the choice of microphone to such things as how to work with many different types of people, how to keep proper and meaningful notes, and giving close thought to the comfort of the artist (the voice actor). Those voiceover recordings that you recognize as setting themselves apart are only achieved through this attention to detail. In music recording, the engineer will be paying attention to whether the musicians are playing in tune and playing on the beat. In voiceover recording, we must pay attention to the quality of the signal coming to us over the monitor speakers, looking for a slurred word, or perhaps an inflection that could be spoken better. And then, of course, we must have the diplomatic skills to gently suggest to the actor that they should do the line differently. As with all aspects of the recording industry, it is the quality of the end product that will make your reputation, and all of the skills discussed in this book are, I believe, essential to arriving at that final product.

This book deals more with techniques of voice recording and only occasionally discusses the technology that we use. There are two reasons for this: First, I am a firm believer that a superior end product can be created with virtually any type of tool (one only has to listen to the classic Count Basie recordings from the 1930s to be convinced of this point of view); and, second, most books that deal with purely technical subjects in the media field tend to become dated quite quickly. Owing to the time it takes to write the book, edit, print and publish, and then finally release and distribute, software changes have inevitably occurred during the process, new technologies are constantly being introduced, and the book is dated before it ever gets to the consumer. I am assuming at least an introductory level of knowledge of computer systems, editing and mixing techniques, and acoustics from the reader who has picked this book up off the shelf. I deal instead with the day-to-day concerns of the recording engineer, producer, and voiceover talent in producing the spoken word. If you feel you need additional guidance with tasks such as editing, processing or mixing, or any other technological aspect of our profession, I would recommend that you do a bit of research and then seek out any of the numerous texts and web sites devoted to these individual subjects. You may find the Resources section at the end of this book of use in searching out further material; many of the resources that I list in this section were very helpful to me in my research for the book.

It is for these reasons that I set out to write this book. Although each voiceover engineer that you talk to will have different approaches and suggestions on how to improve your own work, and some engineers will disagree with my methods and suggestions, be aware that there are many ways to get to the destination, and there is no "right," single way to approach a session. What you now hold is the result of my career in the business, and my experiences are by necessity different from others', but I do believe that a careful reading of this text will give you a good foundation from which to discover your own approach. By diligent practice of your own recording techniques and your own experimentations, you will get to your own, unique way of recording this most important material. But, before you set out on the path to making your own way of doing this, you should be aware of what have been long-accepted ways of doing things. As legendary baseball manager Leo Durocher once said, "I believe in rules—sure I do. If there weren't any rules, how could you break them?"

Tom Blakemore
Chicago, Illinois USA
August, 2014

Acknowledgments

There were a number of people who were instrumental in helping me with this project and who helped bring it to fruition. I'm sure to leave someone out of this mention, and, if that's the case, the oversight is not intentional, and I offer my sincere apology.

At Taylor and Francis, Anais Wheeler was incredibly patient in guiding me through the process in her role as Acquisitions Editor. Her suggestions and gentle humor gave me confidence that the subject matter was worthwhile, and that I could complete the book. Also at Taylor and Francis, Editorial Project Manager Meagan White never seemed to tire of my endless questions and revisions, and her hand-holding throughout the process was reassuring. Her patience at times seemed boundless. And thanks to the Technical Editor, Charles Jude Weinberg, for his comments and insights that kept me on track with all of the myriad technical details this text required.

My appreciation goes to all of the recording engineers, voiceover performers, and producers who took time out of their busy schedules to answer my research questions—all of your input helped inform the direction this book would take, whether your comments are included here in print or not. I would especially like to give a special thanks to three individuals whose help was paramount: Terry Schededer, for his input and suggestions throughout the writing and for his friendship throughout our professional careers (not to mention his grilling recipes); to Harlan Hogan, for sharing his experiences and for helping me to navigate my way through this publishing adventure; and to John Storyk of the Walters–Storyk Design Group for his conversations on acoustics and studio design. I'm not sure I could have completed this without their generous help.

My appreciation to Tribeca Flashpoint Media Arts Academy (TFMAA), for use of their facilities for photography, and to the students, whose curiosity and dedication led directly to the discussions in this book. Also, to my colleagues at TFMAA for suggestions and general conversation on the subject that led to further thought and exploration on my part. As well, my gratitude to Resolution Digital Studios, Second City Sound, Noise Floor Recording, Ltd., and the Chicago Recording Company for use of their facilities for photographic purposes.

I can't leave out a mention all of the copywriters, without whom we wouldn't have any words to say and record.

And, finally, the largest thank you in the world to my best friend, wife, and life partner for her understanding, and for allowing me the time and space to accomplish this. She gave me confidence when I was discouraged and prodded me on when I got lazy. Without her input and trust in me, this book would never have been started. Keeping me on an even keel through all of my many moods, she kept the household together, and for that alone there are not words of praise high enough. So, thank you—and now I promise I'll get back to working on the house! And I can't forget my good buddy Travis for his limited understanding and patience; now we can get back to the park, chasing squirrels again.

Introduction

One microphone, one actor, and one channel on the mixer. What could possibly be simpler or more straightforward? This is how the majority of spoken-word recording is done, and on the surface it appears there's nothing to it. Certainly nothing compared with a high-powered music session, right? Well, if there's one thing that I've learned in more than thirty years in the studio, it's that this can often be the most demanding recording situation that you'll be thrown into. Think about this for a minute: If you are performing with a symphony orchestra, you are playing with perhaps eighty other musicians. But now step onto a stage with only your instrument and with the audience in front of you. There's no place to hide, and every moment is magnified. It's the same with recording the human voice. One microphone, one voice, and only one channel on the mixer. It's now just you and your skill to make or break this moment. Much of the information that we share as human beings comes through spoken words, and the majority of media that we come into contact with every day has a spoken component. Radio, television, movies, video games, even the annoying "Your call is very important to us . . . a representative will be with you shortly"—they all rely heavily on listening to someone pass information to us through words. And the interesting thing to consider is this: Some audio professional somewhere is responsible for recording all of this material.

One microphone, one voice, and only one channel on the mixer. Simple, right? With this book I hope to give you some techniques to think about and put into practice to make recording the spoken word a strong part of your professional repertoire.

A LIFE IN SOUND

For as long as I can remember, from when I was very young, I've been fascinated with the sound of the world happening around me. While my friends were playing baseball or gazing at a line of ants, I would stand transfixed, listening to the sound of a running stream, or to a far-off train. When I was about seven or so, my parents gave me a small transistor radio as a Christmas present. A radio of my own, to listen to whatever I wanted! Late on Christmas afternoon, I lay on my bed and turned on the radio. I always thought that radios were for listening to music or the ballgame, but that afternoon I tuned in to a broadcast of a radio play—*A Christmas Carol*, by Charles Dickens. I had never experienced anything like it—a complete movie playing only in my head and imagination, performed by actors in a studio, using their voices to bring this story to life. On that long-ago Christmas, I discovered the power of the spoken word, and I still marvel at it to this day. Later, in junior high, I acquired a small tape recorder and began recording the sounds of my neighborhood. I soon discovered that I could play these sounds backwards and that, by using a pair of scissors and some Scotch tape, I could put the sounds in a different order. From these childhood beginnings, a life in sound as a recording engineer was born.

Back when record stores were the only way of purchasing new music, every record store had a large section called "Spoken Word Recordings." Comedy albums, audio documentaries, recorded speeches—all of these and more found their way into the spoken-word bins, and in those days sales were brisk. With the demise of vinyl, spoken-word recordings morphed into podcasts, walking tours, news and information on specialized topics, and the like. But the spoken word is still with us, now more than ever, and on a greater range of topics. All of this information must be recorded, edited, and produced, and this is the topic of this book. The greater the skill with which the voice is recorded, the wider the potential audience for this work.

A BIT ABOUT THIS BOOK

This project was conceived at the 2012 AES annual convention in San Francisco. While browsing through the selection of titles available at the Focal Press booth,

I was struck by the absence of any information on the subject of voiceover recording. After a lengthy conversation with the person at the booth, and after thinking about it for the following two days, on the four-hour flight back home I started putting together some notes and a sample table of contents and sent them off to the woman I had met and talked to in San Francisco. It turns out that she was an Acquisitions Editor at Focal and she began to guide me through the process of getting a green light for my idea. Now, here we are in 2015, and the results are in your hands.

In the following pages, I am going to take a few things for granted. First, that you have a working knowledge of basic recording techniques, including microphone design and signal flow in the studio, and that you are familiar with audio editing concepts in general. In the examples, I will be using Pro Tools screens and terminology, but the concepts that I'll be discussing apply equally to any digital audio workstation (DAW) platform, as well as good old analog recording and editing on that quaint medium, magnetic tape. After all, we'll be discussing capturing the sound of the voice and not the theory of any one recording medium. Also, this is not a book that discusses voice acting, recording vocals for music, location recording for film and television, or audio editing and processing in general. Those subjects are deserving of books devoted exclusively to each subject, and indeed there are a number of fine texts available on those topics. But I've found that, for spoken-word recording, there's just not a lot of information out there.

What I *will* be talking about in these pages is improving your techniques in recording voiceover performances for narrations and commercials, documentaries, corporate communications, and games and animation. We will also take a brief look at the world of recording in the field, doing interviews and roundtable discussions. Also, we can't really discuss any of this without revisiting the types of microphone and processing that are incorporated into this work. So, in this book, we'll take a look at the various types of microphone and when you would use one versus another, understanding how the human voice is produced, what makes a good-sounding recording environment, working with voice actors, capturing a high-quality recording of the voice, and specific techniques for various types of media.

And finally, we'll be dealing with the space in which the voiceover reading takes place, the studio, or booth, and not the control room, where the engineer listens to the sound coming from the monitor speakers. My definition of the two might be stated, "studio = performance space" and "control room = recording space." To clear up any misunderstanding, throughout the book I make reference to "the studio"—this is the performance space, whereas "the recording studio" is the business itself, which is comprised of many different rooms—offices, lounge area, tech shop, the control room (or more than one), the studio, storage vault, etc.

The book is structured into two parts. In the first part, Chapters 3 and 4 cover technical and theoretical information on acoustics and microphones that we should be aware of, and Chapters 4–7 discuss the role of the voice engineer and the session itself. Chapter 8 is devoted to designing the voice studio and is aimed primarily at those who would like to record themselves at home, as many voiceover professionals do. The second part, Chapters 9–14, is devoted to discussing specific voiceover situations and types of session (commercials, games and animations, interview situations, and a brief mention of other types of voiceover opportunity). At intervals throughout the book, there are short sections that I am calling "Insight" sections. These offer thoughts on various aspects of our profession, and my guest contributors offer their views gained from many years of experience in the industry. Of course, I am very indebted to these individuals for sharing their thoughts with you. Read them closely and learn from their wisdom.

As I mentioned above, over the course of thirty-five years in the studio, I've been fortunate to encounter a wide variety of recording situations, almost all of them involving working with the human voice in some manner. Drawing on this experience, I would like to share some of these situations with you and give you some small insight into the skills that each requires. It is my hope that, by sharing some of what I've learned with you, you can more easily make your way in the industry and avoid some mistakes that most of us have made, thus becoming a true professional more quickly.

THE MAKING OF A SUPERIOR VOICEOVER RECORDING

I've noticed that those engineers who consistently produce great-sounding voice recordings have a number of things in common. First and foremost, they are world-class multitaskers. When doing any type of recording, we must be aware of many stimuli coming at us at once. With voice recording, we oftentimes must keep our ears open for problems and mistakes in the recording process and, at the same time, keep our eyes and concentration on the script and be on the lookout for missed, slurred, or mispronounced words. Also, we must be aware of the narrative flow and make sure that the tempo, style, and dynamics remain consistent. At the same time, we are documenting the recording and keeping track of timings and good takes, and being aware of possible editing choices we will make later.

Of course, making sure that the actor and the director have fresh coffee, water, and so on is part of the job, as is acting as amateur psychologist to keep the session running smoothly and tempers on an even keel. And we have to keep an eye on the clock to ensure that the session is accomplished in the time allotted and we don't run over and cost the client extra money. All of this adds up to an incredible level of attention to detail that separates the extraordinary engineer from the average, and keeping all of this in mind is how you keep working on a steady basis. So, in a sense, it's as much (or more) about these multitasking skills as it is about your recording skills. Of course, technical knowledge is the bottom line in any job in professional audio, but these skills are what will set you apart from the competition.

If there is any one piece of advice that I can give to anyone doing any type of recording, it would be this: Always be prepared and ready to roll at all times. Digital audio is cheap—if someone is warming up or practicing, why not drop into record? You never know—sometimes that's when that magic take happens, and you don't want to miss it. At the least, you can delete those takes if you want to, and you're not out anything, but miss that one special performance, and it's gone forever. What this means is that, if the client says she wants the session to begin at 10:00 A.M., you have everything set up, the coffee made, the script copied, and a session built in your workstation, and you have verified that the signal path is working properly. At 10:00 A.M., you're all set to press the record button, capture every moment of magic, and come out with a superior product. Do all of this, and the client will return time and again; furthermore, the voice actor will appreciate your efficiency and hard work, and, the next time someone asks him for a recommendation on where to record the clients' new project—well, who do you think that actor is going to suggest?

This book is aimed at anyone who is interested in learning how to effectively record the human voice. I came to learn about this intuitively, during time spent on the job, as I started my career in the studio, but only as I began to study communication theory and pay attention to how the voice is captured did I truly start to understand all that is involved. I'll leave the mathematical proofs of much of what I discuss to others, and, if you're interested in the math of how things work, I advise you to discover a number of excellent texts on the subject, some of which I list in the bibliography for this work.

A few words on terminology: The end product of our work can go by a number of names: "voiceover," "VO," "narration," "read," and so on. I use these terms fairly interchangeably in the book; they all refer to the recorded work. Throughout the book, I'll be referring to both "the microphone" and "the mic." I do this to avoid redundancy; in our world, either is perfectly acceptable. By the way, "mic" is the shortened version of "microphone"—it's never "mike," which is a person's name. Also, when referring to certain people and roles, I use "he" and "she" interchangeably, not only to avoid redundancy and seeming (possibly) sexist, but also to avoid the dreaded "he/she" usage or the creation of some sort of strangely androgynous figure ("the person" or "them," and so on). As well, the person that you are recording can be called "the actor," "voiceover talent," "talent," "narrator," or simply "the VO." All of these refer to the same person and their role in the studio. And finally, whether you are a student, a beginning engineer or a seasoned pro, we're all in this together; we're all recording engineers. We simply have different levels of experience. And so, throughout this book, I refer to "we."

I'm discussing voiceover recording in this book, but much of the material can be applied to any type of recording that you may do.

Whether you are engineering a voice session or recording yourself, there are some things that don't change: The physics of how sound reacts in a room, how a microphone functions, reaching an agreed-upon quality, and, above all, pleasing the client. The better that you understand the technical aspects of the recording chain, the better your final product will be, and the more clients you will have. You'll have to become familiar with software programs, accepted editing practices, and proper mixing techniques. This book will give you some practical advice to start you down this path, as well as introducing you to a variety of markets that need your work and that you may want to explore, to help you make a living in the world of voiceover.

I would like to address two ways of looking at things: Facts and opinions. If I tell you that a particular microphone is a condenser microphone, that is a fact. It can be proven, and there is no doubt about it. Much of the information in this book is opinion, based on my experience in recording the human voice since the late 1970s. Over that time, I've made plenty of mistakes and also made some discoveries that helped me along, and it is these experiences that I would like to pass along to you, so that you can, in a sense, shortcut this process and learn from my mistakes and discoveries in order to get your recordings and your career to a higher level in a shorter period of time. You may have a different viewpoint on the best way to approach voiceover recording, and ten engineers will give you ten different opinions on what makes the best recording situation. All people are different, and what works for me may only hold you and your creativity back. There are no absolutes (except for those pesky facts). Be open and be aware that your opinion is just as valuable as mine and demands equal consideration.

A SHORT CASE STUDY

To help you begin to appreciate the multiple skills needed for effective voiceover work, I offer up this short story about an experience that I had some years back. I was booked to engineer a session at 8:00 A.M. one morning. The afternoon before, I had received a phone call from the producer's assistant, who was calling from England. It seems that a famous British actor was in town filming a movie, and he was going to be recording a 2-minute public service message that would air worldwide. The producer was in London and would be directing the session via phone from there. They obviously wanted top recording quality, and, owing to the demands of the movie schedule, the actor would have only 45 minutes to complete the recording process, because he had to be on set for makeup shortly thereafter. I was to deliver three digital audiotape (DAT) tapes (you can look that antiquated format up on the Internet) via three separate overnight delivery services to the United Kingdom and also archive a safety copy at the studio (this was before the advent of digital phone setups and the Internet, by the way). At the end of that day, the actor was leaving the country, as soon as his shooting ended. It was evident that I had only one shot at this, and the reputation of the studio and my skills were to be tested that morning.

The morning of the session, I got everything set up and tested the signal path, including the phone connection into and out of the console. Scripts were copied, the coffee was brewed, and the studio was set up and ready to go before the scheduled start time. When the actor arrived (hungover, no less), it quickly became clear that he was nervous doing a voice-only recording that demanded emotion and empathy on his part; he had spent his entire career as a film and stage actor and was much more comfortable being able to use his body and facial expressions to convey emotions, rather than just his voice alone. My initial choice of microphone seemed wrong to me after speaking with him for a short period; there was something in his voice quality that I wanted to capture and I recognized that my first choice of mic wasn't the one to use. A quick change of mic and mic positioning and we were off and rolling, watching the clock the entire time. At first, owing to his nervousness, none of the takes were usable, and we pushed on. I did all I could to boost his confidence, as well as supply him with hot coffee in the hopes his hangover would subside. The takes added up.

Well, the session was completed, a composite edit of various takes was made, copies were shipped, the message aired around the world, and no one was the wiser that there were some rather large bumps in the road to getting there. All in a day's work in the voiceover recording business, but not only did I have to know my microphones and the technology that we were using, keep a close watch on the clock, and note the proper takes in order to make the editing process go more smoothly and give the producer what he was expecting to receive, I also had to practice a bit of amateur psychology and diplomacy. As we'll discover in this book, this "bedside manner" is an all-important quality to nurture in yourself and oftentimes makes or breaks the session. One microphone, one actor, and one channel on the mixer. What could possibly be simpler or more straightforward? As you have already begun to see, recording a voiceover is not the time to set up any old microphone, push the record button, and then sit back and read the newspaper.

So, with that, let's take a look at capturing the spoken word—the VO.

The Voice in Media

Human communication is primarily verbal; for tens of thousands of years, our stories, traditions, and knowledge were passed from one generation to the next with the spoken word. Even when the written word is used for communication, the letters on the page are only a *code*, a *symbol* that reflects the language of the individual and the sound of the words. In the modern era, this verbal communication now surrounds us and has taken us back to the oral traditions of our ancestors. Into this world come the art and science of capturing these utterances and passing them along to others. The majority of our learning, interpersonal communication, and entertainment is based on listening to others speak and to the content of their words. We do this instinctively, and we are all quite adept at analyzing the slightest subtlety in what we are hearing, even if we are not consciously aware that we are doing it. We are all expert in picking up the slightest cue in the voice of others and attaching meaning to what we hear. Walter Murch, in his excellent article "Womb Tone" (refer to http://transom.org/?p=6992), informs us that the first of our senses to turn on while we are in the womb is our hearing. Even before we come into the world, we are aware of sound: The pulsing of our mother's blood, the muffled sound of her voice singing to us, and vague, unexplainable sounds from the outside. From the very beginning—from *before* the beginning—we are learning to be experts in the meaning of sound.

Before we delve into the subject of recording the human voice and the tools we use to accomplish this, I believe a little background is in order, to put things into perspective.

INFORMATION THEORY

- 1876—Alexander Graham Bell invents the telephone.
- 1877—The Bell Telephone Company is formed.
- 1877—Thomas Alva Edison invents the phonograph (and the recording industry is born).
- 1920—Regular commercial radio broadcasting begins in the US.
- 1927—*The Jazz Singer*, the first feature film with dialogue, debuts.
- 1948—Claude E. Shannon publishes the first of his papers on communication theory.
- 1980s—The proliferation of digital in the audio industry.

From the first date on this list until the last, there is a common thread running through all of them that can help explain why we record the spoken word the way we do, and the tradition behind it. Both Bell and Shannon were mostly interested in communication by telephone (for the purposes of our discussion here), and this fact is significant. Shannon's communication theory, now known by the more inclusive name of information theory, can help us understand how our spoken communication works. Claude E. Shannon (1916–2001) worked for Bell Laboratories and was, by training and by profession, a mathematician. During World War II, he was involved in the field of cryptology and, in his work, found that the key to "breaking" a code was being able to recognize repetitions. All language contains repetitive sounds, words, spellings, and phrases. For instance, in English, the letter "q" is always followed by "u." Therefore, if a message contained the letter "u," there was a probability that the letter preceding it would be a "q" a certain percentage of the time. From this, he began attacking how a telephone message might be more efficiently delivered. He was working on the problem of how to maximize the distance that a telephone conversation could be transmitted clearly; at the time, there was a significant amount of line loss to the signal as it traveled over distance, and this required multiple amplifiers (also known as repeaters) along the way. Bell Telephone (by then known as AT&T) wondered if there might be a better way to attack the problem. Working on this, Shannon began to develop what he eventually published as his Theory of Communication and, in July of 1948, published his "A Mathematical Theory of

Communication" in the *Bell System Technical Journal*. The second portion of this paper was published by Bell Labs in October of the same year. He started from the beginning: The nature of communication and the problems such communication faced. All signals, all communication, travel the same path—from *source* to *destination*.

No matter the type of signal—written, oral, electronic—they all travel this same path, from source to destination. Shannon termed source and destination *transmitter* and *receiver*, owing to the electronic nature of his work, and these terms are useful to our discussions here.

The telephone message was a continuous signal (much like the recording on an analog tape). The cause of unclear communication, he theorized, was that, between the transmitter and receiver, there was *always* something else, noise. This, he stated, was anything that hindered the clear and unimpeded transmission of the signal (communication). It could be anything, depending on the nature of the communication. If I were to write you a letter, but you had trouble reading my handwriting, this would constitute noise in the communication. Were we to be having a conversation in a restaurant, but the restaurant was crowded and filled with the sound of multiple conversations, we would have trouble clearly understanding each other's words, and thus the meaning would be lost. In a phone conversation, there may be static on the line, and this noise also impedes the transfer of information. Indeed, both distance and the air itself can act as noise elements in spoken communication, making the precise meaning of the words unclear.

Noise is usually thought of as some type of unwanted sound (hums, buzzes, outside traffic sounds, and so forth), but, in the context of information theory, "noise" is defined as anything than hinders the comprehension of the transmitted message, whether that be excessive reverberation, a strong foreign accent, or possibly your listener's mind wandering. Anything that impacts the intelligibility or clarity of the communication can be considered noise. Our goal, then, is to maximize the intelligibility of the message. In discussing Shannon's information theory, Jon Gertner summarizes the problem in this way:

> *All messages, as they traveled from the information source to the destination, faced the problem of noise. This could be the background clatter of a cafeteria, or it could be static (on the radio) or snow (on television). Noise interfered with the accurate delivery of the message. And every channel that carried a message was, to some extent, a noisy channel.*[1]

Think about that again for a minute: All messages, moving from source to destination, must travel across a noisy channel. Our goal in recording the spoken word, then, is to reduce the noise as much as possible and increase the level of understanding at the destination—our audience.

Relating this theoretical discussion to our subject at hand, recording the spoken word, let's change our terminology once more. Instead of "source/destination," which we changed to "transmitter/receiver," we might now substitute "spoken word" and "comprehended word." The theory still applies.

Once again, the meaning is clear: We must do all we can to reduce the noise and increase the intelligibility of the message—the spoken word.

Incidentally, in working on this theory, Claude Shannon did not care what the application of his theory would be. He never considered the practical application of his work, dealing only in the abstract mathematical proofs of his ideas. However, from this paper came the beginnings of digital audio. As mentioned, this work was done with the intention of solving the problem of transmitting telephone messages over long distances. A revolution in how we handle messages was beginning.

---

1 Gertner, Jon, *The Idea Factory; Bell Labs and the Great Age of American Innovation.* New York: Penguin Press, 2012, p. 128.

THE BEGINNING OF DIGITAL AUDIO

Long-distance telephonic communication was plagued by the introduction of noise into the signal as the transmission traveled through the line. During World War II, Shannon was very interested in problems dealing not only with creating coded messages, but also in how to "break" the codes of the enemy. He realized that the more complex the message, the more complex the transmission path must be. Also, he realized that the English language is filled with redundancies, and one of the key methods of breaking a coded message is to search out the redundancies and repeated parts of the message. He sought out a simpler method of encrypting messages that not only avoided redundant words and phrases, but, likewise, avoided the nature of a continuous signal, which telephone messages entailed, and thus would prove to be less likely to be broken. Moving away from thinking in terms of a continuous signal, Shannon began formulating a method of encoding the message as discrete "packets" of information, each of which could be sent long distances without loss and being capable of being decoded at the other end in their original (or near-original) state. To do this, he theorized that the signal could be sampled and mapped, and, instead of the original signal being transmitted, this map could be sent, much more efficiently and without the introduction of the noise element. He postulated sampling the information up to 8,000 times a second (a far cry from our current practice of sampling audio at 48,000, 96,000, or even 192,000 times a second!), and, upon reception, the message could be read back as a continuous signal, without the introduction of noise. By sampling the message, instead of sending the original, continuous information, this process sliced the continuous audio into discrete "snapshots" of that particular moment in time. The message could be transformed from words and phrases into a series of "on/off" instructions, or the 0s and 1s of the binary system. Surprisingly, this method of encoding messages was far from new. One of the earliest methods of long-distance communication was the telegraph, operating with a code made of dashes and dots (the Morse code). A telegraph message can be seen as a binary encryption, and, from this code, words and sentences can be read and thoughts can be transmitted over very long distances. For instance, the phrase "The Beginning of Digital" translates in Morse code as:

$$- \cdot \cdot \cdot \cdot \cdot / - \cdot \cdot \cdot \cdot - \cdot \cdot \cdot \cdot - \cdot \cdot \cdot - \cdot - - \cdot / - -$$
$$- \cdot \cdot - \cdot / - \cdot \cdot \cdot \cdot - - \cdot \cdot \cdot - \cdot - \cdot - \cdot \cdot$$

(The slashes in this example illustrate where each word begins; in sending or receiving a telegraph message, there are slight pauses at word ends. The end of a sentence is noted by tapping out the word "Stop": $\cdot \cdot \cdot - - - - - \cdot - - \cdot$). We see a series of dots and dashes, which are heard on the receiving end of the communication after being transmitted electronically. It is a binary system, just as 0 and 1 represent our current digital binary code. Each of these 0s and 1s Shannon termed "bits"; he didn't coin the term, but was the first to use it publicly. This word is an abbreviation of "binary digit" and was first used by Shannon's colleague at Bell

Labs, John Tukey. Shannon explained that the bit, "corresponds to the information produced when a choice is made from two equally likely possibilities." In binary digital language, "The Beginning of Digital" is represented as:

```
01010100 01101000 01100101 00100000 01100010 01100101 01100111
01101001 01101110 01101110 01101001 01101110 01100111 00100000
01101111 01100110 00100000 01100100 01101001 01100111 01101001
01110100 01100001 01101100
```

Each of these discrete "packets" of information can be reproduced exactly any number of times, without any generational loss and without the introduction of noise. If, for some reason, noise does interfere with the translation of the message, error correction is built into the system to predict which of the missing bits is the culprit, and that bit is restored. The 0s and 1s correspond to the "off/on" voltage of an electrical signal and, thus, are capable of being easily dealt with by computers, which are binary in their operation and operate with "off/on" coding.

Working with Shannon, Harry Nyquist (1889–1976) developed what has come to be known as the Nyquist Theorem, which is essential to the successful encoding of digital information. Simply stated, Nyquist showed that, in order for any frequency to be successfully sampled, that sample must be quantized, or measured, at least twice each cycle, to measure the frequency content of the sample. This is necessary for the decoding of the message to be able to predict the correct frequency content of the sample. If the original, continuous signal was thought of as a "waveform" of discrete frequencies, each being sampled and then represented by a series of 0s and 1s, that waveform could be reproduced exactly at a later time or in a distant place. On the receiving end of the message, the 0s and 1s were used to reconstruct the waveform, and an exact copy of the original message resulted. It is essential that each frequency be represented by two points per cycle, if the frequency is to be accurately reproduced. Thus, if we wish to sample and reproduce a frequency of 1 kHz (1 kiloHertz, or 1,000 cycles per second), the sampling rate must be a minimum of 2 kHz. In Shannon's early work (and the Bell System's early transmissions), he was working with a sampling rate of 8 kHz; the resulting signal could only be reproduced to an upper frequency limit of 4 kHz. So, when we discuss the *sampling frequency* or *sampling rate* in talking about digital audio, we are referring to the upper limit of the frequency range that we can record and reproduce. The *bit depth* of a sample is directly responsible for the dynamic range of our audio—the greater the bit number of a sample (8, 16, 24, and so on), the greater the dynamic range of the recording. Although the digital signal is represented by the 0s and 1s, known as binary digits, or "bits," the sampling rate and the bit depth should not be confused. The practical application of this theorem for our work in recording the human voice will be further explained in later chapters, but it is essential that Nyquist's work be kept in mind when digitizing any audio signal.

The system of digitally encoding the information came from Bell Lab's involvement in the development of a system known as "pulse code modulation," or

PCM. As seen above, the series of dashes and dots, or 0s and 1s, create this "pulse" and can be read as the encoded message. Although not invented at Bell Labs, PCM was perfected there, with Shannon's help.[2] Whether it be a dash or a dot, or an off or an on voltage, these various codes are a matter of "yes/no" choices and can be summed up as follows:

> [1] All communication could be thought of in terms of information; [2] all information could be measured in bits; [3] all measurable bits of information could be thought of, and indeed should be thought of, digitally. This could mean dots and dashes, heads or tails, or the on/off pulses that comprised PCM. Or it could simply be a string of, say, five or six 1s and 0s.[3]

Shannon's "Mathematical Theory of Communication" is largely that, a paper of mathematics, and, if you are interested in this part of our discussion, I urge you to reference any number of excellent books on the subject, some of which I list in the Resources section of this text. Through his work, Shannon showed that, "all information, at least from the view of someone trying to move it from place to place, was the same, whether it was a phone call or a microwave transmission or a television signal."[4]

This pioneering work at Bell Labs was an ongoing effort to improve telephone services, and all communications, and soon led to the transatlantic cable, satellite communication, and the cell phone. The efforts at Bell Labs have important implications for our goal of recording the spoken word for media applications, all thanks to Claude Shannon.

HISTORICAL PRECEDENTS

As vocal media expanded beyond telephone communications, many of the findings that were useful in achieving a higher level of clarity in a telephone conversation became commonplace in new technologies. Early radio broadcasts faced many of the same inherent problems that were found in a telephone call, namely, the introduction of noise into the signal owing to the sound of static and the limited frequency bandwidth available to the early pioneers in these industries. With telephone communications, it was found that, if the bandwidth was limited even further, greater intelligibility of the signal (transmission) could be gained, at the same time reducing the nonessential frequencies of the voice that are not needed for effective communication. Many of our forerunners in the radio sound business came from the telephone industry, because of their experience with voice communication and electronics, and brought their skills with them to the new medium. As a result,

---

2 Gertner, Jon, *The Idea Factory; Bell Labs and the Great Age of American Innovation.* New York: Penguin Press, 2012, p. 126.

3 Ibid., p. 129.

4 Ibid., p. 130.

recordings of early radio broadcasts often sound quite thin, and there isn't a full range of the frequency spectrum in the broadcast. Sacrificing *fidelity* for *clarity* was a tradeoff worth making.

Then, in the mid-1920s, a new drive was begun to bring sound to motion pictures. As film moved into the sound era, the same problems were encountered as found in telephony and radio with respect to the clarity of the human voice. Early movie theatres were usually converted from vaudeville theatres, and there wasn't much thought or energy given to making these venues suitable for playback of film sound. The theatres were somewhat reverberant, because they were built to accommodate musical performance, were not soundproofed, and had little to no acoustical treatment of any kind. Quite unlike the modern film theatre, which is specially constructed for even distribution of the audio signal across the entire frequency spectrum, and facing the additional problems of early amplifier and loudspeaker designs, these theatres were quite poor in reproducing voice recordings. Additionally, the means of recording the soundtrack were limited to two choices: Either recording and reproducing the sound for the film on a disk (the Vitaphone system), which contained static and pops, or recording directly onto the film using a process known as "optical recording." Again, this system was inherently noisy and was not capable of reproducing the entire frequency range of voice and music. Because of these limitations, again we find that there is a boost in the mid frequencies of the dialogue recording to increase understanding of the words, with little to no artificial reverberation added to the soundtrack (which, when added to the natural reverberation of the theatre space, would further blur the sound of the voice). As a result, although the soundtracks of early films may sound thin and have a very limited bandwidth, this was a conscious decision on the part of the sound engineers of the day, to help overcome the limitations of the playback systems.

In the late 1940s, television broadcasting was coming into its own, and this form of communication faced the same problems and limitations as radio of the time, with the added problems of both how the program was heard and where it was heard. Most home living rooms are not quiet places—noise from the outside and sounds created within the home (refrigerator running, forced air heating systems, and so on) mask a significant amount of the sound bandwidth, and living rooms are generally reverberant rooms to a certain extent. Also, the speakers in a typical television receiver were tiny—in the area of only 2 inches, and a speaker of this size cannot adequately reproduce much bass information. As a result, the early television engineers, many of whom had come from the world of radio broadcasting, turned to what had worked for both telephone and radio and began frequency limiting their signal. As a result of all of these technologies using this method of increasing the intelligibility of the signal and overcoming the "noise" portion of the communication chain, as stated by Claude Shannon, we have all come to accept this altered sound as natural and what we expect to hear. By repeated exposure to this way of reproducing the voice, we have become acclimated to the sound and, to this day, accept the boosted midrange as the proper method of experiencing recorded vocal information.

Not all traditions use this midrange boost, however; in the French and English film industries, more emphasis is put on accurately reproducing the sound of the program (more emphasis on fidelity rather than clarity), owing to the history of classical-music broadcasting in early radio in these countries, and, as a result, many of the early films from Europe sound unclear to us in the US today. By the way, I should point out that becoming aware of this factor is important to our understanding of recording the spoken word and will help us achieve a much more pleasing recording; many manufacturers design into their products a midrange boost to help increase intelligibility, and, if we become used to this sound, without being aware of the design decisions made by the makers of our equipment, we are apt to add even more boost to the signal and end up with an overly bright sound. This boost is built into both microphones and loudspeakers, and being aware of these design parameters will help in your final product.

PSYCHOACOUSTICS

The work that Claude Shannon and others at Bell Labs were doing showed one amazing insight: In recording, transmitting, and reproducing any information, the *content* of that information does not matter. Although this may seem perfectly obvious to us now, at the time Shannon published his paper, the idea was earth shattering. The recording, transmission, and reproduction of information are, by and large, technical matters, and we should never lose sight of that fact. We work in a technical field, and this is our main focus, as engineers and as others who want to capture the spoken word for any type of medium. The more we can learn about how the technical aspects of our task work, the more closely we can get to the desired outcome that we hear in our "mind's ear." When working in media, it's obvious that the content of what we are recording matters—after all, that is the point—but how the audio signal (the voice) is generated and how we (and our audience) perceive that signal are equally important, if we are to be successful in imparting the information content to our intended listeners.

The field of psychoacoustics is very complex, and many people spend their entire career studying this subject and formulating new theories of how sound interacts with the human brain, but suffice it to say, for the purposes of our discussion here, that psychoacoustics can be simply defined as how the brain perceives and processes audio stimuli—that is, our subjective response to sound. Understanding how our brains perceive the sounds that we hear is essential to effectively transmitting the information in the message to our audience. (For a discussion of how the human voice is produced, see Chapter 4, Microphones.) By taking into account both how we perceive sound and our knowledge of the intended mode of playback when listening, we can begin to understand how to overcome the noise inherent in any signal and increase the intelligibility, or clarity, of the information content of the recorded signal. It is generally accepted that human beings have a theoretical audio sensitivity of between 20 Hz and 20,000 Hz (20 kHz) in the human audible range.

I say theoretically, because we are all individuals, and all of us have a different range of sensitivity owing to the physical shape of our ears and heads, our gender, our age, the possibility of hearing loss, and other factors. However, if there is equal power across all frequencies that we hear, we do not perceive the loudness of all these frequencies equally, and we can use this fact to our advantage.

Loudness Versus Frequency

If a given frequency is played at a certain SPL (sound pressure level), and we perceive it as being X loud, how strongly must another frequency be played so that we perceive it as being equally loud? It turns out that we don't perceive all frequencies as being equally loud across our audio perception spectrum, even if the SPL is measured as being identical. This is owing to the fact that we are most sensitive to frequencies in the midrange portion of our hearing, around 3 kHz or so. As the frequency is lowered or raised, our sensitivity falls off. This phenomenon can be charted, as seen in Figure 2.1.

Figure 2.1 shows equal loudness curves, also known by the names of the two scientists who developed this theory, Fletcher–Munson curves. What we see is that, for signals of equal strength, the signals at the lowest and highest parts of the graph fall off dramatically toward the threshold of hearing, whereas there is a rise in the perception of loudness in the mid frequencies, centered around 3 kHz. This becomes important to us in the mixing stage of our project especially—to sound equally as loud as the midrange, more acoustic energy is needed at the lowest and highest frequency ranges of our program material; the more bass is added to the content of the program material, the more energy those frequencies take up, although we may not appreciably notice more bass content in the mix. We can see the loudness on a meter, but don't necessarily experience it with our ears.

Perception Versus Frequency

As we evolved as human beings, our hearing adapted to provide greater protection from threats in our environment; as a result, we are quite sensitive to sounds that present the greatest threats and have lost sensitivity to other frequencies. As with the perception of loudness versus frequency, we perceive frequencies in the midrange area with greater sensitivity, with that sensitivity falling off in the upper and lower frequency ranges. As this midrange is where we are most sensitive to sound, it is in these frequencies that we perceive the most information that we receive and, thus, the most clarity and intelligibility.

Thus, when recording and processing the voice for media, we can increase the intelligibility of the voice by increasing the mid frequencies (depending on the voice, generally in the 2–3 kHz range). Care must be taken when adding a boost to these frequencies, or the voice can become thin or nasal sounding, but a small increase can be of value to us. Alternatively, we can reduce frequencies in other bands to increase clarity, particularly those frequencies that lie in the lower spectrum of the

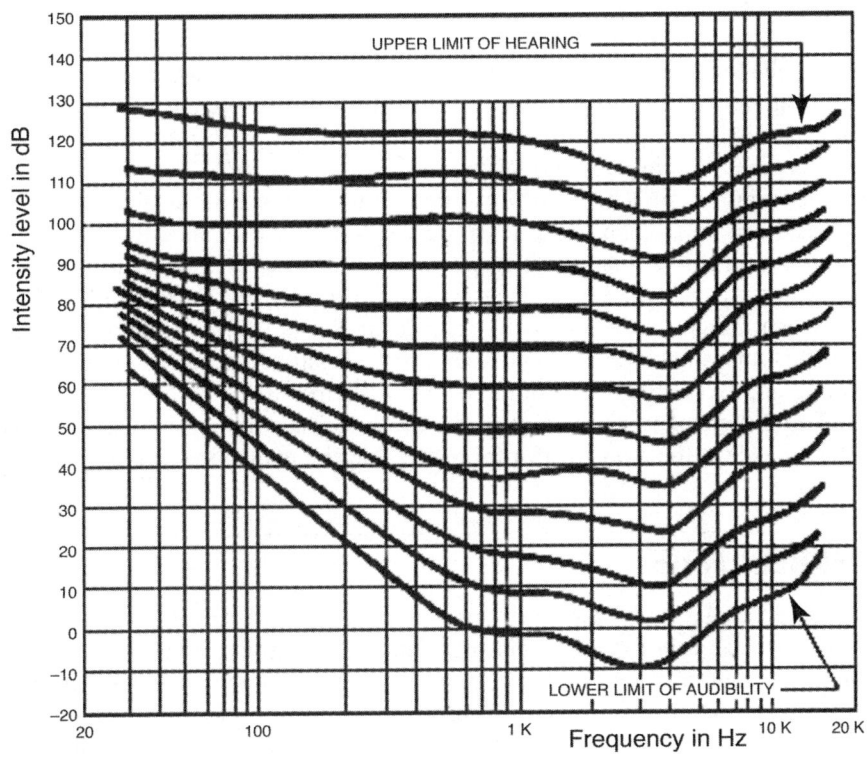

FIGURE 2.1

Equal loudness curves

Source: Courtesy of Walters-Storyk Design Group

human voice. By a reduction in the amount of acoustical information competing for our attention, greater intelligibility is gained. Once again, care must be taken in processing the voice in this manner, or the voice will rapidly become thin and unnatural sounding. There is a fine line between having a natural-sounding recording and one that may carry the greatest amount of clarity but doesn't strike our ear as a natural and full representation of the human voice. By applying these principles of psychoacoustics and how the human brain responds to sound, we can begin to mold our recorded voice to give the audience a greatly enhanced listening experience. For example, if we want to produce a television commercial that stands out from all of the commercials around it, we might boost that 1.5–3 kHz midrange frequency (as well as add some compression to the audio signal to reduce the dynamic range); the resulting recording will sound not only brighter, but louder also. And, our audience will gain more information as a result (see Figure 2.1 above). Another added benefit might be that the voice will tend to stand out from any music we use in the commercial, provided that we also lower those mid frequencies in the music. There are many other purposes for putting this information to work for you, but the idea remains the same in any signal that we hear. These psychoacoustic principles are used routinely in making the voice more intelligible in very noisy environments, such as in radio communication to an airplane pilot or race car driver. The more the mid frequencies are boosted, the more the voice on the radio will stand out from the background noise of engines, rushing air, and other sounds.

The opposite can also be used for dealing with the voice in media: If we were to eliminate these midrange frequencies, we would lose clarity, and, for the purpose of an intentional effect, this might prove very useful to us. Think about the sound of the human voice in the film *Saving Private Ryan* (directed by Steven Spielberg, 1998). When the character played by actor Tom Hanks hears the sound of heavy gunfire, the voices of those soldiers around him are filtered down in the frequency spectrum, approximating the sound of temporarily losing one's hearing. So, manipulating the frequency content of the program material can be used to either heighten or reduce the amount of information that the audience receives.

REVERBERATION

We are able to tell a large number of things about our environment from the sounds we hear: The size of a room, the distance from another sound source, the location of that sound source, and so on. We can use these acoustic cues to great advantage when recording and mixing sound for media. One of the most useful psychoacoustic principles is *localization*—that is, determining the direction of a sound source. We are able to receive sounds from 360° around us; the ability to perceive the sound as coming from, let's say, our left front is mainly a function of our hearing with two ears—the difference in direct sound from the object arriving at one ear and a slightly delayed signal reaching the other ear, owing to both the distance between our two ears and a slight change in amplitude due to the acoustical shadow introduced by the head. This effect can be very effectively utilized when mixing in stereo, but working in a monophonic environment makes this left–right location impossible. If we were to mix in a surround-sound mode, we could also include front–back information to further localize sounds in space. However, even in monophonic, we can quite accurately reproduce a spatial element, at least in two dimensions. By using reverberation, we can approximate the distance of a sound source from the listening position. Thinking about the real world, when someone at a distance speaks to us, we hear more of the reverberation characteristics of the room we are in, as well as a roll-off of high-frequency information. As well, there is a drop in amplitude of the person's voice as a result of distance. So, by combining these three effects, the front–back plane can be convincingly introduced into a monophonic recording. We can see that the three elements needed to introduce a spatial dimension to our work are frequency, amplitude, and time (the reverb element).

Reverberation also can allow us to make a sound fuller; as we add reverberation to a signal, the individual reflections that make up the reverberant field come together in an *additive* manner and reinforce one another. When combined with the original, direct sound, these reflections can provide a richer and more up-front sound. However, if too much reverberation is added, the sound starts to become muddy and hard to define, and so we must be judicious in the amount of reverberation that we add into the original signal. In the case of having too much reverberation, instead of reinforcing our direct signal with the additive effect, we move to the position of the reverberation adding too many discrete signals to what we hear, and thus we can't make out the original signal as clearly. (For a further discussion of the effects of reverberation, refer to Chapter 3, Room Acoustics.)

THE DISEMBODIED VOICE

In working with the human voice in any type of medium, there are two choices that the actor and the engineer, working in concert, can make in deciding on an approach to the performance and the recording. These are what I term "the personal voice" and "the disembodied voice." In most media applications, we deal with the personal voice—that is, the feeling that the actor is speaking directly to each one of us, in a personal way. Think of a line from a commercial: "Let me tell you about this wonderful new dish soap that I tried from Sudso Soaps." Or perhaps this line from a political spot: "A vote for Congressman Moneygrubber is a vote for the special interests, and that's not the American way." In both of these instances, the voice is speaking directly to each of us individually, one on one. The voice is recorded relatively free of reverberation and in a dead-sounding room, and the microphone is close to the actor's mouth. The resulting intimate sound places the listener's ear close to the actor, and the result is that of hearing a friend giving you a piece of advice or information. The actor and the engineer together make the choice of how these lines will be recorded—one with the delivery and closeness to the mic, and the other with the type of microphone and the processing employed in the recording and mixing of the voice. Note that this personal approach does not necessarily depend on the volume at which the lines are delivered; it is more a question of style of delivery and recording. The personal voice is a matter of someone having a conversation with you, regardless of the medium; at times, our conversations become boisterous, other times they are quiet and contemplative, and at times we are simply stating something to the listener. But all of these situations are examples of that personal voice.

Another use for this approach is the narration that we often hear in film (both documentary and narrative) and games. The opening narration spoken by Tommy Lee Jones in *No Country for Old Men* (directed by Joel and Ethan Coen, 2007) is a perfect example of this type of delivery. Recorded at close proximity to the actor's mouth, with little to no reverberation and sounding brighter than his lines on screen (and recorded on the set), this opening narration is Tommy Lee Jones's character speaking directly to each one of us and informing us of his state of mind and the situation he finds himself in. There are numerous examples of this type of voice recording in film. These narrations provide the listener with background information and let us know more about the character speaking the lines. In video games, this technique is used for the same reasons, to great effect. And, of course, there is the movie cliché of hearing the character's voice but not seeing their lips move, while a noticeable amount of added reverberation is also added: Ah, the "hearing their inner thoughts" scenario. Effective? Yes. Original? Not so much.

For those performing the voiceover, remember this: The voiceover is usually thought of as a one-on-one conversation; the actor isn't a rock star shouting, "Hello, Cleveland!" in a stadium. Rather, the voiceover is about holding a conversation with a friend, even if we are aware that millions of people will hear the words. And it is this intimacy that we are trying to capture, with all of its friendliness and personal content. Don't *announce*; have a *conversation* with a specific individual. The voiceover is speaking to one person and telling her why you believe in what is being said. Remember, she won't believe it if you don't. It is this intimacy and speaking to the individual that give the voiceover its power and appeal. Don't announce—talk; tell a story, even if that story is about a new laundry detergent. Remember, the microphone is your substitute for your listener's ear. One person is your audience when you are using the personal voice. Those voice actors who are the least effective are those who use the disembodied voice, instead of the personal.

At the other end of the spectrum is the disembodied voice, often referred to as "the voice of God." Having less to do with the deep, rumbling voice of an almighty, it is the voice speaking to us from above. The difference in the approach is that, whereas the personal voice speaks to each individual, the disembodied voice speaks to all at once: The same approach that one would take when speaking to one person, or addressing an entire room full of people. The disembodied voice is omniscient: It comes from everywhere and from nowhere at the same time. We hear this approach many times in announcements; for example, at a sales meeting, we might hear, "Ladies and gentlemen, please welcome the President and CEO, John Smith." The disembodied voice addresses a large number of people at the same time. In this instance, the recording would be done with the microphone at a greater distance from the actor (this may be a subtle difference in placement), and the equalization would not be as bright. There may be a small amount of reverb added to the voice, but be careful in adding reverberation to this type of delivery—if the room that the announcement is to be played back in has its own reverberant character, the added reverb may lead to a loss of intelligibility of the voice recording, and the message could be lost. Also, if the intended playback venue is likely to be crowded or noisy, such as at an airport, added reverb will add to the loss of clarity and understanding that the listener can gain from hearing the announcement. Even though the voice is generally recorded from a slightly further distance from the actor, we're still concerned with clarity in the recording. The disembodied voice generally uses more projection on the part of the actor while speaking. So, the difference between the personal and the disembodied voice is in the approach taken in both delivery and recording.

Of these two approaches, the personal voice is the one most often encountered in media applications. We are trying to create the illusion of information flowing from one person to another, even if the delivery method is a national television commercial that is experienced by millions of people at the same time. The voice in media is mainly used for passing information; in a video game, the voice is often used for character development, the health status of the character, background information, warnings, and other purposes. In a telephone prerecorded message, we might learn how much we still have in our savings account at the bank, set up an appointment with our auto mechanic, or learn how to update our stock portfolio. We might encounter an online tutorial on how to effectively use a new piece of software, and so on. In each of these instances, information is being delivered from one person to another for education, information, entertainment, or all at the same time. At the root of any recording of the spoken word is the choice of the personal or disembodied approach. As the engineer, you play a large part in determining how the audience perceives the approach.

RECORDING FOREIGN LANGUAGES

As a voiceover engineer, you may be requested to record various projects in a foreign language. This is nothing to shy away from and can add to your client base, even if you are not fluent in the language(s) that will be recorded. In fact, this work can be quite lucrative; for instance, in the world of video games, there is a process known as "localization." The idea is that the player of the game wants to hear the dialogue in his or her own language. Unlike in movies, subtitles aren't very effective in games, and, as a result, all of the dialogue must be translated into various other languages. This is a large and growing field of recording work that you may want to explore. In carrying out this aspect of VO recording, I have discovered that, far from it being an intimidating situation, the recording of foreign languages can be quite educational for the engineer in ways you may not have thought about.

A few years ago, I was in France, and late one night decided to see what was on French television at that time of day. I came across a nature documentary that was produced in Germany, then dubbed into Italian and now broadcast with French subtitles. It was one of the more interesting audio experiences that I've come across: Because I don't speak any of those languages with any facility, I couldn't understand what the narrator was saying. As a consequence, there was no content to what I was experiencing, only pure audio. I could concentrate solely on the quality of the narration recording, the effectiveness of the sound effects, and how the mix was done, without being distracted by the story that the documentary told. It had something to do with birds in a forest, but that's all I could get out of it. The key here was that there was no meaning gained from the *content* of the program. Try this experiment, which I do with my sound design students: Watch a film that is in a language you do not understand, but that has good sound design, and disable the subtitles on the DVD. You will find that you can concentrate on the sound much more effectively because you're not being distracted with the story and with trying to figure out who the characters are and what the situation is. The quality of the audio becomes much more apparent, because your brain is desperately trying to make sense out of what you are experiencing. With no words (content), all of the sound becomes much more important in this process of trying to decode what is going on. All you are left with is pure sound. This can be a very powerful way to study the way sound works in film and is pertinent to our discussions of recording voiceover as well. (By the way, the film I usually do this experiment with is the 1981 German film *Das Boot*, directed by Wolfgang Petersen, although any film with good sound design, in any language you don't speak, would work just as well.)

If we can divorce our brain from the language component of what we are hearing, we can focus much more deeply on the pure sound of what we are hearing. Our brains are wired to search for content and meaning when we hear a voice, but, if we can't understand the language in which the voice is speaking, we ignore the built-in multitasking that we do all the time, without even realizing, when analyzing and taking meaning from the words, and we can concentrate much more closely on the *quality* of the voice. This is the state that good voiceover engineers strive to reach in all of our sessions, but of course it becomes very difficult, because we are also supposed to be listening closely for mispronunciations and grammatical errors while we're recording. We search for meaning even when we are having trouble making out what the message is. Hearing a foreign language is encountering a code, and, if you remember what Claude Shannon discovered in his work in cryptology during World War II, the key to deciphering a code lies in recognizing repetitions. This became very apparent to me during the course of a multiday session, recording the voiceover for a series of training videos on the maintenance and repair of road-building and construction vehicles for a major construction industry supplier, for use in the Japanese market. I don't understand Japanese, and Asian languages at times present a particular problem because of the way that words are inflected at the end of sentences; these languages don't follow the European model of ending a sentence on a down inflection—quite often, the word ending that ends a sentence in Japanese or Chinese (either Cantonese or Mandarin) is inflected upwards. As a result, I have trouble spotting the sentence breaks in these languages, and editing becomes very difficult. However, in doing these Japanese-language sessions, I found a very interesting thought process developing.

The name of the manufacturer was the same in either English or Japanese, and I could pick that word out of any sentence quite easily. During the course of the session, I began to notice that the product name was oftentimes followed by a word I didn't understand, but, in referring to my English script, I saw that, in English, the product was followed by a certain word. I asked the Japanese translator who was producing the session if I had made the correct translation, and it turned out that I had. I began to look for other patterns like this, and by the end of the recording and editing found that I was beginning to learn a small amount of Japanese because of the experience. Now, this doesn't prove my facility with foreign languages, but rather that recording in another language is simply part of the job and something that can be important in learning our craft. When recording and editing languages of Western European origin, I usually don't have any problem whatsoever, even if

I don't speak that language. The reason is because these languages use the Roman alphabet, as does English, and so, even if I might not understand the meaning of the words, I can follow along on the script and know where I am, and I'm able to mark my edit points and later make those edits without any problem. The scripts that are impossible for me to deal with are those that do not use the Roman alphabet, but rather some other form of written communication, and for these sessions I need a translator sitting next to me telling me where to make the edits. Russian, Chinese, Japanese, Arabic, Hebrew—all are a complete mystery to me. But here's the thing: The less I understand of the words (the content), the more I can pay attention to the technical quality of the recording. If we can train ourselves to reach this point, even in dealing with a script that's in our native tongue, we will become better engineers. Learning to divorce yourself from the content, even when keeping in mind that paying close attention to content is a vital piece of the job, is a skill worth developing and one that marks a good engineer. Granted, it takes a splitting of the mind to be able to do this, but it can be accomplished. Don't back away from recording in another language—it can be fun and educational and open new doors for you as a successful recording professional. All of the rules of VO recording still apply, and the same equipment is used. After all, a human voice is a human voice.

If you have the time and skill, consider learning another language. Where I live and work, the most common foreign language spoken is Spanish; where you live it might be something else. By becoming bilingual, you will become immensely more employable and open up new markets for your studio owner, especially if the studio owner is you!

Coming back to where this chapter opened, beginning with information (or communication) theory and from there understanding how the human brain perceives sound impulses, we can start to recognize the essentials of vocal communication and how our recording of the spoken word fits into these patterns. By putting these principles into action when recording the voiceover, our projects will be better able to reflect the communication intended and help our listeners more easily grasp the intent and meaning of the words.

Room Acoustics

Achieving superior quality in the recorded voiceover is a result of the interaction of three factors: The room, the voice, and the microphone. In the following two chapters, we'll look at each of these. This chapter deals with the studio space and acoustics; Chapter 4 will delve into how speech is produced, and how the microphone captures the spoken word. Please be aware that this chapter is not meant to be a comprehensive examination of the subject of acoustics, but rather an introduction to the subject for the purpose of achieving a good-sounding recording in the studio.

The first step in evaluating the studio space is to *listen* to what the room sounds like; get out the diagnostic tools and measure all you want, but simply listening to what the room is doing can get you far down the road to solving potential problem areas. The other thing to keep in mind is that any evaluation of the studio space must be done in that room itself. All rooms will have their own acoustic signature, and that includes the control room—just like the studio, the control room will have its own reflections, reverberation, and other anomalies. Judging the sound of the studio while monitoring in the control room will introduce the control room's acoustics to the studio's, and it will be nearly impossible to determine accurately what the studio sounds like. Monitoring on headphones will eliminate the control room's signature and let you determine what the studio's acoustics are doing. The listening space can influence our perception of the sound, and so acoustic control in all areas is essential.

With a voice recording, our goal is to capture as pure a signal as possible and to eliminate any noise factors that we can. "Noise" has been defined as "unorganized sound," and, for the purposes of this discussion, we can say it is any extraneous sound that is not the spoken word. It can be classified as external sound getting into the studio, excess reverberation in the room, sounds from loud HVAC, and the like. For the voice recording, we want maximum intelligibility and clarity in the recording; any effect that we would like the finished and produced recording to have we normally add in the editing and mixing stages. Above all, intelligibility is the one absolute that we strive to attain. Anything that intrudes upon the recording will distract the listener, and intelligibility will be diminished. And keep in mind that, unlike with a musical vocal, there is often little or nothing that will mask these unwanted sounds (music or sound effects are most times played at a much lower level than the voice to retain the intelligibility).

To achieve this, we must first review some fundamentals of acoustics and the way in which sound is propagated in an enclosed space and then captured by the microphone. This chapter will look at the voice studio—the physical space itself, how the microphone "hears" the space, and some thoughts on setting up the studio to maximize the recorded quality of the voice. There are four factors to consider when attempting to control noise (or other unwanted signals) from intruding into the spoken-word recording: Isolation, room dimensions, the room shape, and the room acoustics. We'll begin with a short discussion of how sound reacts in an enclosed space—the studio.

When a sound is created, the sound wave radiates out from the source in all directions, until it either dies away or is reflected by a surface.

Sound waves react to reflection from a surface in the same way that light waves do; that is, when a sound strikes a surface, it is reflected at the same angle that it originally struck at. This pattern of reflected sound is known as the "angle of incidence" and the "angle of reflection." The sound wave then travels through the air until it strikes another surface, where it bounces off that wall, losing energy and becoming softer with each reflection, and so on, until it dies away and is no longer heard. Another way of thinking about this and visualizing how sound interacts with a room is to imagine a billiard table. The table has no pockets for the ball to disappear into, and wherever the ball strikes one of the sides of the table, it will rebound back. So now we roll a ball across the table (all of this is assuming that there is no spin or "English" on the ball that would affect the angle at which the ball rebounds), and it bounces off the side. Notice that the angle of reflection equals the angle of incidence, just like with the analogy of a beam of light above. If we could see all the trails that the rebounding ball makes as it travels around the table, we could begin to visualize what happens to sound in an enclosed space. To help control these reflections, we can add something that absorbs some of the energy of the ball; in this case, we can replace the hard rubber of the table sides with a softer foam. The foam will absorb some of the energy of the ball, and it will lose momentum on the reflection. In the same way, we can add absorption to a room with acoustical treatments and reduce the amount of reflected sound. Now, let's change our billiard table to a pool table and introduce pockets into the corners—when the ball goes in this direction, it disappears into the pocket, and there is no reflection at all. This would be analogous to bass trapping in the corners of the room.

Of course, sound isn't a single-point event like a beam of light or a ball rolling across a table; it radiates in all directions at the same time. If we imagine an infinite number of billiard balls all emanating from the same source point, radiating in all directions at the same time, and then reflecting off of all six boundaries, we can begin to see the complexity of how sound acts in a room. It becomes imperative to control these reflections, and this is the aim of the acoustic designer.

Whether we think of sound waves as acting like light waves or billiard balls, in neither case does the wave travel straight out and disappear if it is confined in a reflective space; it always bounces at the angle of reflection. Besides the six surfaces that make up the room—four walls, ceiling, and floor—it is wise to remember that anything within the room will also reflect sound waves, to a greater or lesser extent, depending on its size. As you can imagine, the wave pattern in a room quickly becomes very complex.

The sound wave is made of molecules of air that are compressed and in relaxation (known as compression and rarefaction). When these areas of compression strike a surface, they are reflected back into the space, where they pile up with other areas of compression. As you can imagine, the resulting sound may not be ideal. One result is that one frequency can cancel another, owing to the phase relationships of the waves, and can result in a comb filter effect and, thus, highly color the sound. What we are trying to do with the acoustics of the voice studio (or any room for that matter) is control these reflections so that they don't color the sound we are attempting to capture. We do this by means of *absorption* and *diffusion*. If we can absorb the sound wave in some manner, it won't have the chance to reflect back into the space and color our sound. Likewise, if we can break the sound wave up so that its angle of reflection does not equal the angle of incidence, it is less likely to present difficulties. All of the above assumes the wave to be a pure sine wave of fixed frequency and is useful in discussing how sound reacts in an enclosed space, but this is never the case, except under test conditions in the studio. The problem in a real-world setting is that how a wave behaves is a result of the frequency of the sound and the resulting length of the sound wave, as well as amplitude and the angles of reflection. For instance, low-frequency sounds have much longer wavelengths than high-frequency ones. As a result, the amount of absorption that must be accomplished is likewise much greater, and a more massive absorber must be used to be effective. This is most effectively done with the use of *bass traps*.

FIGURE 3.1

Absorber Panels

Source: Courtesy of Auralex Acoustics

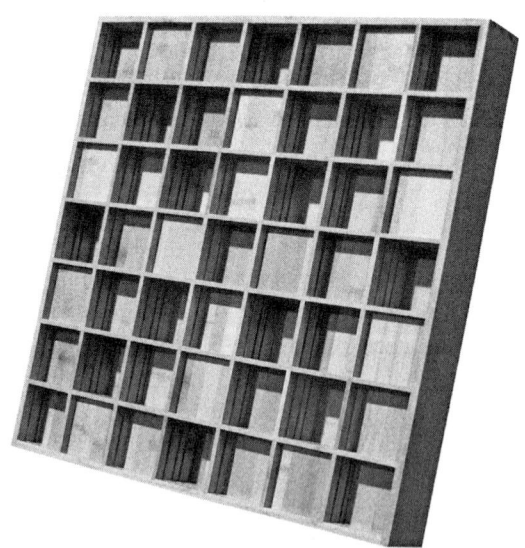

FIGURE 3.2

Diffuser Panel

Source: Courtesy of Auralex Acoustics

Sound reflected off walls, ceilings, and floors directly influences the quality of the sound picked up by the microphone. At the microphone, the majority of energy captured will be *direct sound*, emanating directly from the source (in our case, the actor's mouth). However, reflected sound also makes its way back to the microphone and interacts with the direct sound, causing many unwanted artifacts. After reflecting off a wall or floor, a signal will bounce back to the listening position. This is known as an *early reflection*. Early reflections help us to determine the size of a room, as well as distance from the source. Sound that reflects off two or more surfaces before arriving at the listening position is known as a *secondary reflection*. As the signal continues to reflect off boundaries, the number of reflections builds until the room is filled with reflected sound, and we now experience *reverberation*, the state in which it becomes impossible to determine the direction of the source sound and which lasts after the original signal has ceased to be emitted. In this instance, the acoustical energy has become equal across the entire room.

The role of acoustic design in the studio space is to control these artifacts and eliminate as many of them as possible. When designing a studio space for music recording, some reverberation is oftentimes built into the room—for this type of recording, the warmth and naturalness of room reverberation is more pleasing than artificial reverb, and, depending on what type of music is being recorded, the decay time of the room is taken into consideration. For instance, a longer reverberation time is often specified for studios recording classical music, so that the instruments avoid sounding "brittle" and "dry." Acoustic jazz may call for shorter-duration decay times, and so on. For voiceover work, the room should be nearly unnaturally dead. The voice studio is the most acoustically dead sounding of any recording space.

We must do all we can to capture only the sounds on which we intentionally want the audience to focus, and this means eliminating as many of the reflections as we can. By absorbing and diffusing the sound waves, we can arrive at much better control of the sound picked up by the microphone and a more pleasing recording.

Part of this control comes from the size of the room itself. I have recorded in all manner and size of rooms, and it is my experience that, the smaller the room, generally the more problems with reflection and coloration you will end up with. This is because the boundaries of the room are much closer to the microphone, making the reflected sound much stronger in relation to the direct sound that the mic is picking up. Also, the shape of the room is of vital importance to the end quality of the recording. I realize that we can't always construct the ideal room and must live with the existing space that we have. The good news is that virtually any room can be improved with proper application of acoustic materials. The first step in eliminating unwanted reflection is to eliminate any 90° angle that you possibly can. A 90° angle (either two walls or wall–floor or ceiling–wall) will lead to a buildup in bass frequencies that must be controlled, either by trapping these frequencies or by eliminating the 90° angle. When constructing a space, take care that there are no parallel wall surfaces. As you can imagine, a sound wave striking a wall straight on will reflect off that wall straight back toward the opposite wall, creating what is known as a "standing wave." In rooms that have a problem with standing waves, you can easily hear the result: A sine wave is played back into the room, and, as you move about the space, the amplitude of the sound gets louder and softer as you move. What you are hearing is the "piling up" of the reflected waves in certain parts of the room (or, to be more accurate, the reinforced compression portion of the waves and the associated relaxed, or rarified, portion), and this will give a highly inaccurate representation of the original sound. Using measurement devices, we can often detect a wide variation in amplitude in areas just a few inches apart. Whether a particular room is a candidate for standing waves can be easily computed: If the roundtrip path length from one wall to the opposite wall and back (front to back or side to side) is exactly the same length as the wavelength of the sound wave, or a multiple of that number, without proper treatment the room will cause a standing wave at that frequency, and you (and your microphone) will hear either louder or softer reproduction of that frequency, depending on where the hearing apparatus is located. Even with a good deal of absorption, it is difficult, if not impossible, to eliminate the resonance. The answer is to avoid the use of parallel surfaces when at all possible.

A room will not absorb all frequencies equally, and there will be a variance in the time it takes for any given frequency to die away; this results in a different reverberation time for two differing frequencies in the same space. These variations

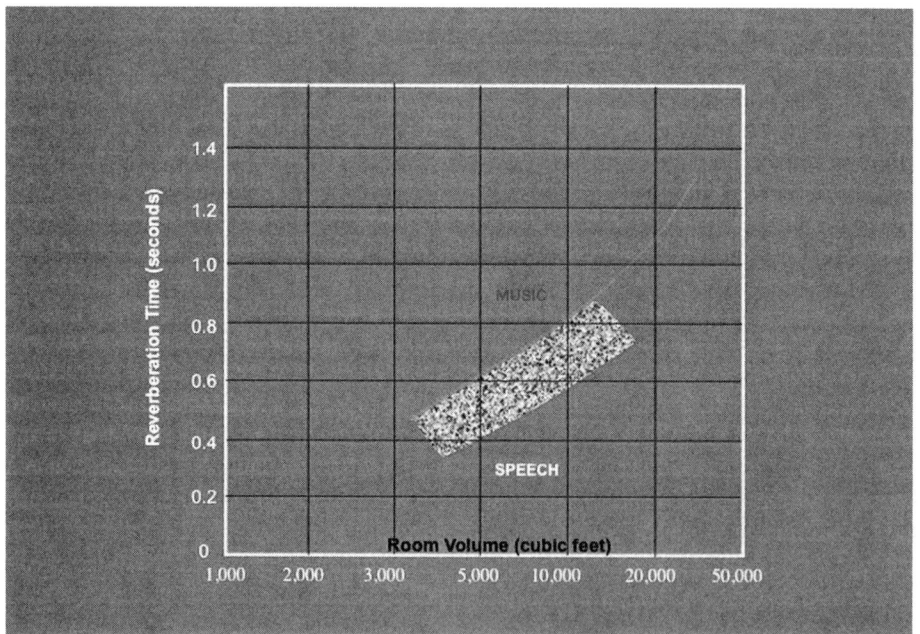

FIGURE 3.3

RT60 Target Values

Source: Courtesy of Walters–Storyk Design Group

INSIGHT: Daniel Porter, Regional Manager and Product Training Specialist, Auralex® Acoustics, Inc.

A Word on Project Studio Acoustics

Now that all you need to produce music is a laptop, the will to create, and a space to work in, the demand for larger traditional studios with acoustically dedicated spaces has been significantly reduced, the obvious reason for this being simple economics. The operating cost of a multiroomed traditional studio can be daunting for both engineers and musicians alike. The majority of music production now takes place in what the industry has labeled "project studios." These types of studio are multipurpose facilities that are housed in a single room and are designed to accommodate every stage of production. More often than not, project studios are found in residential areas and are converted living spaces of some sort (bedroom, office, closet, etc.). The rapid development of project studios by both hobbyists and professionals is a direct result of the development of affordable and easy-to-use digital recording technologies. The following is a list of five fundamental acoustical topics that are applicable to nearly all project-studio environments:

Isolation (Room-to-Room Sound Leakage)

Unfortunately, the only way to truly isolate a studio is to add multiple layers of mass with varying densities to the existing walls, floor, and ceiling, or to build an entirely new space. This often proves to be too costly an option for project-studio operators. For those without the money or the permission of the landlord to alter their studio so dramatically, here is a helpful tip: Think of sound like you would light or water: If you were to fill your room with water, where would it leak out? If you shone a spotlight around the room, where would it escape to the exterior?

Addressing these weak points in your room's inherent design with off-the-shelf weather stripping or homemade MDF window plugs can produce audible results. This will give you better insight into how isolated the studio is already, and what steps need to be taken to improve it.

Acoustical Symmetry (Studio Layout)

When choosing a room to convert into a project studio, the rectangular design that dominates living spaces is often the best option, despite its parallel surfaces. Even though such spaces are havens for flutter echo and low-frequency summation/cancellation, they allow for one to orient the listening position and studio monitors symmetrically with respect to the sidewalls. This ensures that, with or without acoustical treatment, the sound in the room is arriving at the listening position symmetrically with respect to the left and right stereo channels.

Absorption (Acoustical Foam, Fiberglass, Recycled Cotton, etc.)

Nearly all project studios acoustically benefit from the application of absorptive materials to otherwise reflective walls and ceilings. Once a room is oriented symmetrically, placing treatment to the left, right, in front, behind, and above the monitors and listening position will reduce the extraneous reflections that would otherwise distort the sound being reproduced.

Bass Trapping (Low-Frequency Absorption)

It is almost impossible to over-bass trap a project studio. However, one can apply so much bass trapping to a room that otherwise usable space is sacrificed for relatively small acoustical gains and substantial monetary investment. For this reason, a compromise in acoustical design is typical. The most efficient place in a project studio for bass trapping is in the upper vertical and horizontal corners. This is where the traps can have the greatest acoustical impact, without limiting the amount of physical space available for production equipment, furniture, engineers, and musicians.

Diffusion (High-Frequency Scattering)

Diffusors are often used as a counterbalance to absorption. By design, they add acoustical character rather than reduce it, as absorption would do. Whether or not this is appropriate is completely at the discretion of the studio operator's needs and wants. In any case, an important thing to consider with regard to diffusion is the distance from the listening position. Keeping most diffusors at least 6 feet away from the listening position will allow for them to operate as they are intended. It is not uncommon to see diffusors on the ceiling or the wall behind the listening position in a project studio. These configurations can create a more natural sense of spaciousness for either recording or listening when used strategically.

For more information regarding acoustic treatment or studio design, visit www.auralex.com

are what give any room its sonic signature, and the room can be "tuned" by the designer to achieve a certain sound and feel. The reverberation time is measured using RT60—the time it takes for a level to drop 60 dB. As mentioned, all rooms and all frequencies exhibit differing RT60 times, and a well-designed room will take this into consideration. Through the use of high-frequency and low-frequency absorption, the RT60 time can be molded and controlled to a great degree. Figure 3.3 shows the ideal RT60 measurements for studios designed for music and voice applications.

FIGURE 3.4

Corner Bass Trap

Source: Courtesy of Auralex Acoustics

When it comes to very large rooms, such as music studios or film stages, the effect can be equally disconcerting. The problem of unwanted reflections is eliminated to a large degree because of the distance the sound waves must travel; however, we now often have the problem of reverberation to deal with. Music studios are oftentimes purposely constructed to have a reverberant quality to a greater or lesser degree, and this reverberation, although accepted and even welcomed in music, is not conducive to a good-quality voice recording. To overcome the problems of a very large room, we can work closer to the microphone, which has ramifications that we should be aware of and that will be discussed in later chapters, or use movable baffles, also known as "gobos" or "flats." These portable panels are constructed using an absorption material such as fiberglass insulation or foam and covered with an acoustically transparent fabric (much like a loudspeaker grill cover). This effectively isolates the microphone from the larger room, creating a smaller and more controlled recording area. These panels are available commercially or are an easy do-it-yourself project if you are handy with basic woodworking tools. By arranging these gobos around a microphone, we in essence create a micro-recording environment within the studio space and can then more accurately control the reflection of sound waves. The gobo can also be used to help isolate one microphone from another acoustically, if more than one actor is recording at the same time.

FIGURE 3.5

Tube Traps

Source: Courtesy of Applied Sciences Corporation (ASC)

If the resources to specially construct the perfect room are lacking, there are prefab solutions available that can be custom ordered in various sizes and shapes and offer many window and door options. A product known as WhisperRoom™, from WhisperRoom, Inc. (www.whisperroom.com), comes in a wide variety of sizes and can be equipped with customizable acoustic treatment, as well as a wide assortment of window and door options. Another product is the Porta-Booth™ (available at www.voiceoveressentials.com). This product, developed by a well-known voiceover actor, encloses the microphone in an acoustic treatment and renders virtually any space quite usable for voice recording. It is (as the name suggests) portable and can be used with a wide variety of microphone types and sizes, in virtually any setting.

Whatever the final decision on arriving at an acoustically pleasing space, there is one aspect to the studio that must not be overlooked, and that is proper ventilation. This is a vital design consideration for two reasons: First, the recording space must remain comfortable for the actor for long periods of time, and the lack of air movement and excessive heat that come with improper ventilation can be very wearing on the person involved. Second, the sound of noisy ventilation can and will ruin an otherwise good recording. Air in motion coming from heating and cooling ducts as well as the fans can generate a lot of noise, and proper precautions must be taken to eliminate most, if not all, of this sound. The ideal solution is to employ baffled ductwork and to use large-diameter ducting and vents. Low-flow air handlers are ideal, as a large volume of air moving slowly is much quieter than a fast-moving amount of the same volume. Designing proper ventilation into the studio upon construction is preferable

FIGURE 3.6

WhisperRoom

Source: Courtesy of WhisperRoom, Inc.

FIGURE 3.7

PortaBooth Pro

Source: Courtesy of Harlan Hogan's VoiceOver Essentials

to attempting to retrofit the ducting and air handling into an existing space, as quite often this would require demolition of the ceiling and some of the walls. For prefab spaces such as the WhisperRoom™, ventilation can be included in the package as an option and is very well designed. If care is not taken in this area, the sound of the ventilation will be very apparent in the recording and will be extremely difficult to remove. Equalization rarely does much good on this type of noise, because the frequency spectrum of the HVAC noise is broadband and includes the same frequencies as the voice, so that attenuating the noise signal attenuates the voice frequencies as well. One other aspect of having the studio space adequately ventilated is that of the life span of the microphone. You don't want to spend significant amounts of money on a mic, only to have it sit in an overheated and humid environment, or one that might have significant temperature variations. We will discuss other considerations of studio comfort and utility in Chapter 6, The Studio (talent comfort, lighting, copy stands, etc.).

There are many other potential sources of noise creation in the studio; fluorescent lights are notorious for their buzzing sound, loose items on shelving can vibrate in sympathy with certain frequencies, computer fans, etc. All of these noise sources within the studio must be addressed, or they will be heard by the audience. The background noise level in a studio (or any room) can be measured using a rating system known as the noise criteria (NC) level. For any intended use of a space, an optimal NC level can be determined by the room designer or acoustician (see Table 3.1).

Besides acoustical problems within the room (reflections, reverberation, HVAC, fluorescent light buzzes, and so on), we also must be aware that the studio should be free of any ambient or external noise. Noise from the outside is constant and can come from traffic, airplanes, weather, and so on. The amount of this outside sound that enters the studio, as well as sound generated inside the room, contributes to the noise floor of the room, and this ambient noise will interfere with the listening experience. The background noise level in a room can be measured and is rated as an NC number.

Table 3.1 Typical NC Ratings

Condition	NC Level
Sleeping	25–35
Living space	35–45
Office	30–45
Restaurant	35–50
Home theatre	20–25
Concert hall	20–25
Motion picture theatre	25–30
Audio studio	15–20

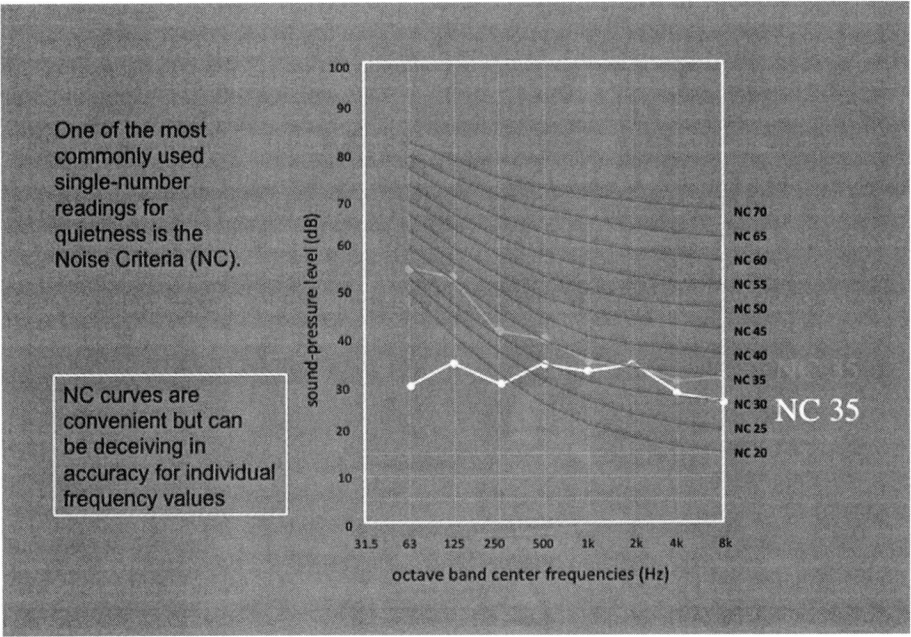

One of the most commonly used single-number readings for quietness is the Noise Criteria (NC).

NC curves are convenient but can be deceiving in accuracy for individual frequency values

FIGURE 3.8

NC Curves

Source: Courtesy of Walters–Storyk Design Group

Depending on the intended use of a given room, varying NC ratings are acceptable: A sports stadium would be around NC50, a movie theatre generally around NC20. For voice studios, the recommended level is around NC12–15, meaning that these rooms are very quiet. If they are not, the noise will not only interfere with the aesthetic experience of listening to the voice, it can become so distracting that the intelligibility of the voice is lost. With voice recording, as previously stated, there is oftentimes nothing to mask this noise on the voice recording.

Exterior noise can come from either of two sources: Air-borne sounds or structure-borne sounds. If there is sound coming from outside the studio that enters through the walls or because of air leaks in the room, that sound will be picked up by the microphone and be a source of distraction to the listener, thus contaminating our clean recording. The same is true of sound that enters through the building structure. These sounds can be extremely difficult to deal with, and eliminating structure-borne noise takes careful planning of the studio space. Most times, this type of noise contamination will be low-frequency information that travels through the building's structure and can be caused by a passing truck or bus or something of that sort. I once was invited to open a new recording space and went to check it out. It was in the basement of a building that was suitable in all regards for a studio space, but, as I stood in the unfinished space, I realized that about every 5 minutes there was a very audible rumbling sound. I discovered that the building was about a block or so from a major subway line, and, as an underground train passed, its vibrations were traveling through the ground and into the building's foundations and then into the air inside the room. Needless to say, I passed on building a studio in that space and moved on with my life.

These types of problem can best be overcome by using what is known as "box-in-a-box" construction. Leave the existing walls, ceiling, and floor as is, insulate all around the interior, and construct a new room that is "floating" within the existing room. By placing the floor of the new studio room on isolating pads or springs and hanging the ceiling in the same manner, then making sure that the new walls do not physically connect with the existing walls, you decouple the new space from the original, and no structure- or air-borne sound is able to enter.

Many of the pre-fab solutions discussed above are fairly simple fixes to these types of noise issue, and, if you must place a studio in an existing space, adjacent sound issues and noise sources should be investigated. Of course, if you begin all-new construction, instead of trying to retrofit a new studio into an existing space, you are free to design in all the proper features, such as non-parallel walls, a floating structure, and so on.

INSIGHT: John Storyk, R.A., Principal—Walters–Storyk Design Group

Studio Design and Acoustics

To me, a studio recording space means any environment with a live microphone. I recognize that recording takes place without live mics, but that is another book. Now, having said that, people do put live microphones in their control rooms, but for this discussion we'll assume there is a control room and there is then one space or X number of other spaces that become the preferred live-microphone locations—in other words *the iso booth*. As always, good design starts with good programming. As designers, the first thing that we try to get our arms around in the studio space is how is it programmed? What type of recording will happen in that space? Music, spoken word, what are we trying to record there? Very quickly in this process, we need to determine the size of this small room—how many people do we need to accommodate? The recording type drives style and acoustic signature. These are the early questions we need to resolve. Almost immediately following this discussion, we need to address issues such as budget, location, ergonomics, natural daylight, special features, isolation, etc. It's a big list, and at times it seems to be a puzzle that cannot be easily pieced together. But it all starts with the type of recording, the type of music or non-music (e.g., voice only), and size. If we don't get these issues sorted out, nothing else is going to make much sense.

If it's voice only, we typically are not as concerned with very low-frequency response issues. Usually, there isn't much recorded information below 80 Hz. For instance, recently we were asked to improve the acoustics of a very small (22 square feet) vocal booth. In this case, as in many booths, there was a "boominess," a lack of clarity at the lower frequencies, that did in fact seem to translate into some of their recordings. The geometry was so small there was no opportunity to really address 80 Hz as a fundamental eigentone (standing-wave frequency). There wasn't even any room for thin membrane absorbers. Our solution in this case was a simple one—try to put some air behind our broadband absorbers, essentially extending the absorptive frequency range of the panels. The recording engineer will accomplish the rest of this task with roll-off equalization.

We're very strict on the NC15 rule for voiceover rooms. Generally speaking, those rooms are on the dry side. Music is much harder to deal with, as it depends on a broader range of frequencies. Even more complicated are multipurpose environments. The fully isolated or fully decoupled room design is typically how an NC15 is accomplished (see NC curves graphic—Figure 3.8). We should remind

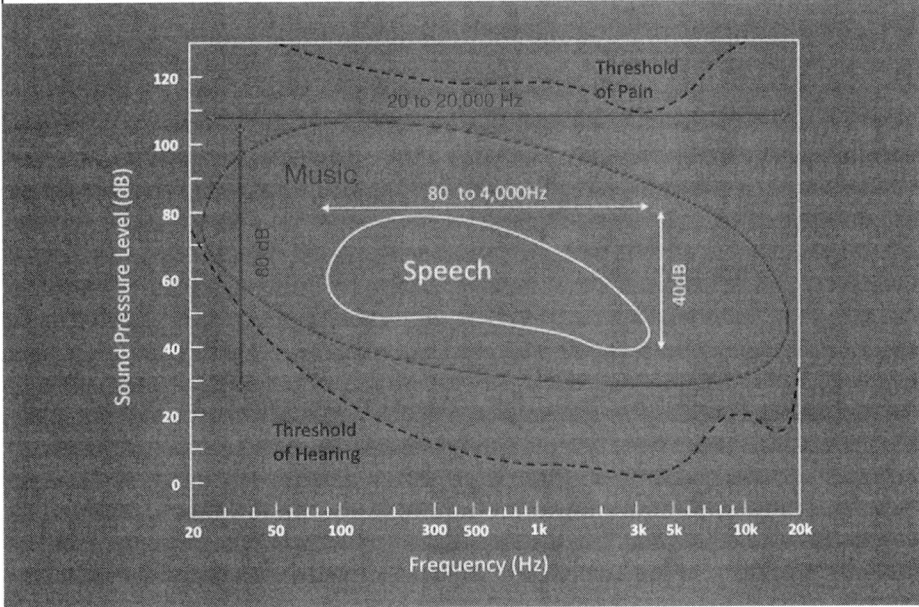

FIGURE 3.9

Relative Sound Pressure Levels

Source: Courtesy of Walters–Storyk Design Group

ourselves of one of the oldest rules of noise control—*to obtain a quiet room, choose a quiet site!*

If you're building an iso booth in the middle of the woods and you don't record when it rains, it is conceivable that the boundary for this room could literally be a tent! Sounds silly, but it's true. More typically, iso booths will tend to be fully or partially decoupled, because they're often adjacent to control rooms, which get loud. Again, these small, quiet rooms need to maintain NC15 quietness values. Decoupling rooms is expensive—no two ways about it. If we're going to make a compromise for budget on a project (this is called value engineering), one can allow control rooms to be a bit louder, particularly larger control rooms that monitor at loud levels. Designing a control room to NC15 is usually a waste of money. I have seen them as high as NC25 and still working very well. Background noise from people and equipment will override NC15 specs!

As far as achieving maximum voice intelligibility, just make sure the room is attenuated as equally as possible through its useful frequency recording range.

The ideal RT60 time for a booth is very low—often .1–.3 seconds (max) at 1 K. Some people would argue that these rooms are so small and so "dead" that they simply do not have a measureable reverberation value. Regardless of whether you think there is reverberation or not, or whether you think it's statistically correct, there is a decay rate. You can measure it and you can define it. For me, more important than an exact value at 1K is the variation between octave bands of this decay rate—particularly as we approach the lower frequencies. In the problematic room we previously discussed, the simple explanation for this poor performance was basically inadequate low-frequency absorption. The "run away" low end was thus easily explained. The room displaying a .2 second decay rate at 1 K, while at the same time displaying a decay rate of nearly 1 second at 125 K, is most definitely not going to perform correctly. Why is this happening? Simple—there's no absorption at 60 Hz in the room. There wasn't any room, or, more typically, somebody just installed foam, thinking that, "You walk in and it sounds dead—thus better." For speech, it's possible this is satisfactory. But, put a bass guitar in that room and it will be bothersome. In general, less foam, which is really mid-frequency absorption, and more attention to low-frequency attenuation might be a better approach to these rooms. Corner bass trapping (low-frequency control) may often be your only option. And, by the way, there's more than just one corner. Most people think of corners in a square or rectangular plan, but you can put the corner traps at the intersection of the wall and the ceiling.

Most projects that start out as "speech-only vocal" booths aren't really, in fact, speech-only vocal booths. In the end, the client really wants an iso booth that can also double for solo instruments—i.e., put a guitar in there, maybe a sax, throw a bass guitar amp in, etc. Many times, a guitar amp is in the small iso booth, and the player is in the control room. So, in an initial client discussion, they will start off with, "Yeah, we're only going to do vocals," and, about 20 minutes later, after you have a little conversation with them, you realize that what you're really building is not a vocal booth but an iso booth. The difference to me is that an iso booth is one that will accommodate music *as well as* vocals. Again, our argument for *planning* and *programming*!

Typically, I would encourage a studio owner to make the booth as large as possible, as this will help tackle the low-frequency issue. For instance, if someone says, "I want an iso booth and I want it big enough for a baby grand piano or big enough for a drum kit," then the footprint of the piano or the drum kit starts to drive the design—simple as that! If it's literally, "I want an iso booth and what I'm really going to be doing there, day in and day out, are one to two vocals but maybe an instrumentalist," then it can be a bit smaller. It always circles back to programming: What are you trying to do in this room? Imagine a room where a baby grand piano has to move in and out on a daily basis. When this happens, we would try for double doors. This is a classic example of an ergonomic condition driving the architecture and ultimately directing the final acoustic solution. Architecture and acoustics—a love affair that leans on creativity and compromise!

For more information, visit www.wsdg.com

Through all of this discussion, the most important point to take away is this: The spoken voice is a relatively quiet source, and any type of noise contamination will be immediately apparent and must be avoided at all times. Unlike music being recorded, the voice at a conversational level is generally in the 65 dB range, and so it is easily disturbed with any type of intruding information, whether that be reverberation, heavy HVAC sound, or the sound of a passing train!

Microphones

"What microphone should I use to record a voiceover?" This is the most common question on this subject that I hear, and the short answer is, "I can't answer that." Sad, but true. It's a bit like you asking me what car you should buy; the answer is, "It depends." Are you looking for a sports car, a pickup truck to use at a construction site, a family van? What is your budget? There are too many choices and too many possible uses when choosing a new car. So, examine your needs and your personal driving style and start whittling the list down a bit. Choosing an ideal, one-size-fits-all mic is much the same—impossible to find. But, in this chapter, maybe we can help you with the decision-making process.

In this chapter, we will examine our tools of the trade—microphones. We'll take a look at different microphone types and what choices we might make to achieve a quality voice recording. The wide range of microphone types and designs can be confusing at first, and, in this chapter, a number of different types of mic will be explored and their suitability for voice recording discussed. We've looked at the acoustics of the room and the importance of taking acoustics into account when setting up an optimal signal chain, and now we will be discussing the next step in that chain and the importance of being aware of the characteristics of various types of microphone. But, before we get to that, let's first examine the instrument, the human voice, and how sound is produced.

FUNDAMENTALS OF THE HUMAN VOICE

Certainly, one of the first things that we must note in discussing the human voice and the recording process is that each of us, whether we realize it or not, is an expert in the subject of the voice. After all, we spend most of our waking hours listening to speech in all its different forms and means of transmission. If we stop and analyze how much information we absorb throughout the day by means of the spoken word, we come to see that any anomaly or distortion of the voice is immediately noticeable and will strike us as being "wrong," without our even stopping to think about it. We can easily distinguish between a "friendly" voice and an "angry" voice; between the authorative or the soothing; between nervous and confident; male and female. Each of us is so accustomed to hearing and analyzing human speech that it has become second nature to us, and we go about our daily lives hardly being aware that we are doing so. So, it is with this in mind that the discussion of the voice, and the microphones used to record it, begins.

Although each of us has a unique voice, and although we may speak in different languages, the method by which the voice is produced is universal. Involving more than just the vocal cords, the sound of speech is produced by nearly the entire upper half of the body, and understanding how the voice is formed is essential to recording it well. There are four areas in the body that provide the sound of the voice: The chest, the throat, the mouth, and the nose. The chest acts as a resonating chamber and is the source of much of the bass content in the voice. As well, the chest is the location of the diaphragm, a muscle in the lower abdomen that pushes upwards on the lungs and is responsible for the controlled use of the breath when one is speaking or singing. As air exits the lungs, it passes into the throat and across the larynx, or vocal cords. Remember that sound is air in motion, and that what we perceive as sound is the vibration of an object that creates waves in the air. It is the vibration of the larynx that creates the sound of the voice.

The modulated vibration of the air column then moves upward into the mouth and the nose. The mouth functions to form the various components of speech and articulate the words with the tongue and lips. The nose is the source of much of the air content that we hear with speech. Each of these parts of the anatomy is important in forming the voice, and each must be considered when recording. The more we understand how the voice is produced, the better the decisions we can make on proper microphone placement and on avoiding the unwanted artifacts that the speaking voice contains. These can include popped Ps, sibilance, and nose honks (yes, a real problem for some speakers). These artifacts and how to deal with them in the course of a session are a main component of our job.

The individual components of speech are known as "phonemes." These are the most basic sounds that we make, even below the syllable level. Knowledge of how the voice is produced is essential in avoiding problems in our recordings. There are certain phonemes, or sounds that we make, that adversely affect the recording when picked up with a microphone, and the most common and problematic of these are called "plosives." These are those sounds that produce an explosive burst of air coming out of the mouth, sounds such as P, D, T. If you speak a word that starts with one of these letters and hold your hand in front of your mouth, you will feel the burst of air, and, if this hits the mic, a loud "thump" will be heard, ruining an otherwise perfect recording. This sound, known as a "pop," is the sound of distortion as the mic overloads from the air pressure hitting the capsule. If you experience this situation, there are two ways of dealing with it: You can reposition the microphone so that the plosive isn't spoken directly into the element of the mic (usually off to one side of the mouth or slightly above, so the air burst doesn't go directly into the front of the microphone), or you can use a windscreen or pop filter. I much prefer a pop filter over the foam windscreen that comes with almost all microphones when you purchase them; in my experience, the foam windscreen, although effective in

FIGURE 4.1

Pop Filter

Source: Courtesy of Resolution Digital Studios; photo by the author

eliminating pops, reduces the high-frequency content of the signal. On the other hand, pop filters, either nylon fabric or a metal grid, are acoustically transparent but highly effective in eliminating the dreaded "P pop." Also, the metal screen is easier to disinfect and keep clean. This is not as inconsequential as you might think—early in my career, I had a job as a radio announcer, and the mics we used had foam windscreens. If one of the announcers caught a cold, it was passed to every other announcer, because of the germs being captured in that foam! So please, be kind to our voiceover announcer brethren and keep those filters clean.

Another sound that will intrude on a perfect take is "sibilance." This is the unnatural stressing of the S sound (or phoneme). Again, this is a type of distortion that we must strive to eliminate. Some people are more sibilant than others, and this can be especially true for women announcers. Metal and foam pop filters are of little use in eliminating excessive sibilance, and your only recourse is a repositioning of the mic. I've found that positioning the mic above the mouth—at about eye level and angled downward—often helps alleviate this problem. There are processors known as "de-essers" that can help with the problem; a de-esser is a frequency-dependent compressor. You set the frequency that you would like to control (the approximate frequency of the offending S) and the amount of cut (in decibels) that you would like, and you can attack the problem in that way. However, being aware of the problem so that it can be rectified at the source is much more preferable than resorting to electronic manipulation. It may take some experimentation with mic placement, but this is by far the best way to go about this problem.

With music, the frequency range that we hear can be quite wide, from an average of around 40 Hz upwards to 20 kHz and beyond. Therefore, choosing a microphone that is capable of capturing this full range of frequencies is paramount. However, the human voice is much more limited in its range: For the average human voice, the fundamental frequency is approximately 125 Hz for males and 225 Hz for females. I say approximately, because all of us have a slightly different pitch to our voice, and this fundamental frequency will shift depending on the individual speaking. The upper range of the voice is in the 1,500–2,500 Hz range for the harmonics of the voice (this refers to the speaking voice—a female operatic soprano singer reaches around 6 kHz). The voice contains overtones above this fundamental frequency.

In music, the dynamic levels are quite wide, reaching perhaps 90 dB SPL or so, but the voice, at a normal conversational level, is in the range of 65 dB. What this means in actual practice is that the levels in a voice studio never reach those found in a music studio, and the acoustic footprint of the room is much less noticeable. However, paradoxically, owing to the relatively quiet nature of speech, the acoustics become all the more important. If there is a significant noise floor in the studio, there isn't enough volume from the source signal to mask the noise. Also, although some reverberation is expected in a music studio, any reverberation in the voice studio becomes immediately obvious, and all steps must be taken to eliminate it. Refer back to the NC ratings in Table 3.1, in the previous chapter, to see how this works in practice. Knowing this information is important, and it will be affected by where a person is placed in regards to the microphone.

MICROPHONE DESIGN TYPES

A microphone is a *transducer*, which is a device that changes one type of energy into another type of energy, in this case acoustical energy (air in motion and the associated sound waves) into electrical energy. There are three basic design types that have been implemented to accomplish this, and each has its own characteristic. First, we have the *dynamic*, or moving coil, microphone. With this design, a thin diaphragm of metal or Mylar is attached to a coil of wire and is set inside a magnetic housing; when the diaphragm is struck by the moving air produced by a sound, the diaphragm and coil move back and forth over the magnet, creating the electrical energy needed for recording. Dynamic microphones are known for their ruggedness in a wide range of situations, and this is by far the most common microphone type for live-performance vocals and field use by news crews and the like. Dynamic microphones often exhibit more self-noise than either of the other two types of design, and they can be limited in high-frequency response and transient response owing to the inertia of the moving coil. Dynamic mics utilize a diaphragm of the largest mass of any of these three microphone types, and this is what introduces the inertia and limits some of their overall response.

Another design is the *condenser* microphone. With the condenser mic, acoustical energy once again strikes a thin metal diaphragm suspended inside the mic. Unlike in a dynamic mic, however, the diaphragm is placed in front of a fixed metal plate that has an electrical charge applied to it. As the diaphragm moves back and forth in relation to the fixed plate, the output varies in relation to the distance of the two elements from each other, and this varies the output signal. So, instead of creating its own electrical signal, as the moving coil design does, a condenser has to have a source of electricity supplied to it to charge the metal plate. This charge is most commonly +48 V, DC, and is referred to as *phantom power*, so called because an outside source (either the microphone preamplifier or a separate power supply) provides the electricity directly through the mic cable, which also sends the mic's output to the mic preamp. Power for this mic can also be supplied by a battery placed inside the microphone's body. Condenser microphones are most often associated with in-studio use, because they can be quite fragile, but their sound is much fuller and more robust than the dynamic's, with a wider frequency range. The electrical output of a condenser microphone is very small, and, therefore, it has to be boosted to be of use, and so there is a small amplifier built into the mic body. This amplifier can either operate with a vacuum tube or using solid state electronics; solid state models are by far the commonest that you will see, although there are tube condensers available, either vintage or new designs, and these are considered to be warmer in sound and more responsive; they are thus often sought out for digital recording because of the "hard" edge that digital recording can introduce. Because of the design, suspending the diaphragm in front of a charged plate, a condenser mic is much more sensitive to changes in air pressure and can pick up subtleties that the dynamic does not; a condenser has far superior transient response and is more sensitive to minute changes in pressure at the diaphragm. Part of the appeal of condenser mics, especially the *large diaphragm* variety, is the richness and fullness that are produced because of this design.

Finally, the third common type of mic that you might encounter is the *ribbon mic*. Here, a thin, corrugated metal diaphragm is suspended between two poles of a magnet, which produces its own electricity. The ribbon mic is a variant of the dynamic design. Ribbon mics are highly susceptible to wind (breath), and not much air movement is needed to knock the ribbon out of the magnet's influence and destroy the microphone. Most older designs are also very susceptible to high SPLs (very loud sounds), which could destroy the microphone, although more modern ribbon mic designs exhibit a much higher tolerance of high SPL values. Highly fragile, they should be treated with due caution. A pop filter or windscreen is a necessity with this type of mic, and be aware that, if you connect a ribbon mic to phantom power, you might destroy the microphone, so always be aware of what you are connecting the mic to before plugging it in. There have been some very interesting developments in the design of ribbon mics in recent years, involving varying the polar response and being able to tolerate very high SPL levels. Also, many of the more recent innovations utilize phantom power for the mics, so, when using a ribbon mic, be very aware of what model you have in your hand. As mentioned, if the mic is not designed for phantom power, you risk destroying it if phantom power is applied. Manufacturers such as Royer, Beyerdynamics, Shure, and others have been making inroads with new designs, but, to date, nearly all are aimed at music recording. Ribbon microphones were, at one time (that is, in the dark ages of early radio), the mic of choice and fell out of favor quite some time ago for use as a voice mic. Once thin and "vintage sounding," the new mics on the market produce a very pleasing and natural sound that many find quite appealing for voice and music recording. Ribbon mics have a lowered high-frequency response curve, and this can give them a warming effect on the voice that many find quite desirable. Ribbon mics have a very low electrical output (the lowest of the three major microphone designs, and, therefore, a transformer must be used to bring the output signal up to a usable level). However, self-generated noise is also quite low.

There are two other microphone designs that we will touch on in this chapter: The USB microphone and the new models of digital mics; however, neither is yet widely used for voice recording. For professional voiceover recording, condenser mics are the most widely used, although, as we will see, certain dynamics are highly regarded as well. Microphones can cost anywhere from under $100 to well over $12,000 for a mint, vintage tube condenser, and so there is a choice and a model for every budget.

POLAR RESPONSE

Every microphone has a pattern of accepting sound that the manufacturer builds into it. Many microphones (most commonly the condenser mics) have a selector switch on the microphone body that allows you to alter this pattern. This pattern of acceptance of the incoming sound signal is known as the *polar pattern* or *polar response* of the microphone. The polar response is a plot of how much of the incoming signal is picked up by the microphone's front, as seen from directly overhead (much as if you were looking at a map of the world from directly over the North Pole), sides, and back. There are five main patterns: *Omnidirectional, cardioid, hyper-cardioid, super-cardioid,* and *bidirectional* (or *figure 8*). Each of the patterns has its own use and function when the voice is being recorded. Every microphone is delivered with a specification sheet that lists not only its frequency response and sensitivity to sound (measured in SPL), but its polar response as well.

Many microphones have a selector switch on the body of the mic that allows you to change the pickup polar response. This feature will help you adjust the microphone to meet a large number of recording situations and room acoustic signatures, and save you the expense of having to purchase multiple microphones. The microphone in Figure 4.3 has five response patterns from which to choose; most mics will allow for perhaps three different patterns (such as omnidirectional, cardioid, and figure 8). Not all mics have a selector switch, so, if this is something that you might find useful, be sure to check for this feature before you purchase.

An omnidirectional microphone picks up sound equally from all directions. Because of this pickup pattern, the omnidirectional mic can not only pick up the wanted

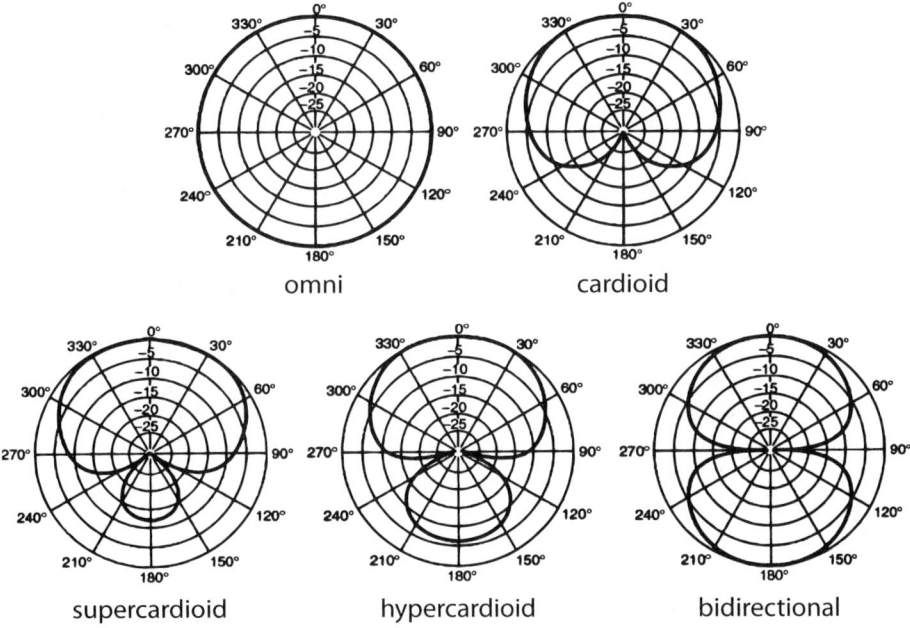

FIGURE 4.2

Polar Response Curves

FIGURE 4.3

AKG C414 Polar Pattern Switch

Source: Tribeca Flashpoint Media AA; photo by the author

signal (the narrator's voice), but also any reflections and room noise present. Although often used in music recording, omnidirectional microphones aren't used with regularity for voiceover work, because of this factor. It should be noted, however, that some broadcast organizations prefer the warm, even sound of the omnidirectional microphone, including the BBC. For recording in a well-designed acoustical space, the omnidirectional mic can have a very pleasing and smooth, warm sound. However, you should pay attention to any unwanted artifacts making their way into your recording if the room is less than optimal. Also, if you must record a number of actors at the same time and only have, or only have room for, one microphone, an omnidirectional pattern may work well. For this reason, they are also used when recording radio dramas and the like.

The cardioid pattern, as shown in Figure 4.2, accepts sound mainly from the front of the microphone, in a shape that is reminiscent of a heart shape—hence the term *cardioid*, meaning heart. Instead of being omnidirectional, this type of mic is *unidirectional*. This is by far the most prevalent pickup pattern for recording the voice. Because of the pattern, the mic rejects sound from the rear of the microphone, eliminating many of the reverberations and reflections within the room, and the mic can be situated to point away from any noise-generating problems. The cardioid pattern lends a phenomenon known as *proximity effect* to the microphone, which is a boot of the low frequencies in the voice as the sound source gets closer to the microphone. This phenomenon can be very beneficial when recording voiceover to help give a larger, "voice of God" quality to the actor and also can help to compensate for a thin or high-pitched voice. However, if the actor has a naturally deep and resonant voice, the proximity effect can lend a "boominess" to the voice and can be distracting to the audience. Proximity effect is most pronounced in ribbon mics and virtually nonexistent in omnidirectional-pattern microphones. Cardioid patterns can be found in dynamic or condenser microphones, depending on the manufacturer's design. Dynamic mics designed for stage vocals are meant to be used quite close to the mouth and count on the proximity effect for their sound. When used at a normal speaking distance, these dynamics often sound thin, without much bass reproduction.

Hyper-cardioid microphones have a pickup pattern similar to the cardioid, but "squeezed," resulting in lower sensitivity from the rear and sides of the mic. This results in a more directional pickup when the sound source is directly in front of the microphone. Sometimes referred to as a longer "reach," and used in shotgun mics, hyper-cardioid microphones don't actually have a "longer" range of acceptance, but do reject more of the sound from the rear and sides. When using this microphone for voiceover recording, the actor must be very conscious not to move his head from side to side when speaking, or the resulting sound will vary greatly and be quite noticeable. For this reason, I would recommend using a hyper-cardioid mic only on experienced voice talent who have a good working knowledge of microphone technique and the technical limitations of various microphones. However, having said that, I can also mention that the voice recorded with this type of microphone is very present and "in your face" and is quite useful in advertising situations especially. Because of this up-front sound, it may prove tiring to listen to in a longer narration read. It isn't as smooth and warm as the sound of a cardioid mic, and using this mic isn't for all voices that you will record. Experimentation is the key here: I would recommend setting up a hyper-cardioid mic directly in front of the talent and another, cardioid, microphone in close proximity, either to the side or perhaps above the hyper-cardioid. Then, by comparing the two signals, you have a choice of which to use.

The super-cardioid design is much like the hyper-cardioid and lies between the cardiod and hyper-cardiod in its pickup pattern. For this reason, side-to-side motion from the sound source can be very pronounced and should be monitored closely. When the hyper- or super-cardioid microphone is used, the directionality of the pickup pattern is achieved through the use of ports along the body of the microphone, called *rejection ports*. As sound enters these holes in the mic body, it arrives at the capsule out of phase and, through the principle of phase cancellation, is not introduced into the recorded sound. Care should be taken never to block these rejection ports, with the hand for example, if holding the mic in the hand, or perhaps with tape or some other foreign substance. This will negate the directionality of the microphone, and more off-axis sound will be recorded. Because of their inherent directional pickup patterns, both hyper- and super-cardioid microphones can be useful in avoiding any source of noise generation in the studio—*if* you are aware of any off-axis sound due to side-to-side movement from the actor.

Then there is the figure-8 pattern, or *bidirectional* microphone. As its name implies, sound is accepted from both the front and the rear of the mic capsule, whereas sound coming from the sides is rejected. Useful for recording two people with one microphone, the figure 8 has the roundness and warmth associated with an omnidirectional mic, and also its inherent problem of picking up sound reflections from the rear. Once again, if choosing this pattern, be sure that the room is acoustically adequate.

FREQUENCY RESPONSE

Not all microphones are designed equally; depending on the mic's design, they will exhibit various responses in regard to the audio frequencies that they "hear." If you were to graph the range of frequencies that any particular microphone picks up, you might notice that, at the extreme upper and lower frequency ranges, there is a fall off in the mic's sensitivity. There may be peaks or a gentle rise at certain frequencies, or perhaps a bit of a dip at a particular frequency.

The curve shown in Figure 4.4 tells us that the frequency response is fairly uniform in the upper bass and midrange frequencies; in other words, we can say that this microphone has a "flat" response. There is a drop off at the upper and lower ends of the scale, and these frequencies lie outside the frequency range of the human speaking voice. As the graph becomes less flat, more colorization is introduced into the sound that the microphone reproduces. This may be a good thing, or it may be an undesirable trait of the particular microphone; it depends what your goals are when choosing a particular microphone. For this reason, in addition to the pickup

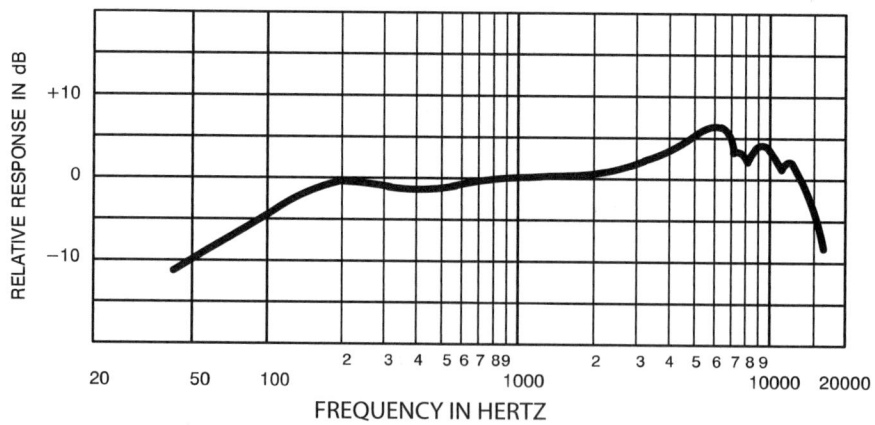

FIGURE 4.4

Frequency Response Graph

Source: Shure, Inc.

or polar pattern of the mic we might choose, we also have to take its frequency response into consideration. By paying attention to the frequency response of the microphone, you can begin to shape your sound without the use of equalization after the fact. If you want a brighter sound, there are microphones that have this built into them; the same is true for a warmer sound, or a sound that is more pleasing for male or female voices. Attention to the frequency response of any given microphone is the first consideration in choosing one over another. There are microphones designed specifically for certain applications—for the bass drum, for instance, or for harmonicas. Know your application and choose your microphone accordingly.

Many condenser microphones have a selector switch on the body that reduces the mic's sensitivity in the low frequencies. Called a *bass roll-off* switch, this feature is useful for reducing unwanted frequencies such as room rumble caused by HVAC noise or perhaps unwanted proximity effect. The bass roll-off can be engaged if you are recording outdoors, to help control wind sound to a certain extent. The bass roll-off attenuates the low frequencies and is a built-in high pass filter.

In Figure 4.5, we see a microphone with a variable bass roll-off switch. This assortment of frequencies from which to choose is unusual, but, as you might imagine, this allows for great flexibility in shaping exactly the sound you want, for a wide range of possible room or other acoustic problems.

Also seen in Figure 4.5 is another selector switch that allows the user to set the microphone's sensitivity to loud signals. Again, most mics that you encounter will

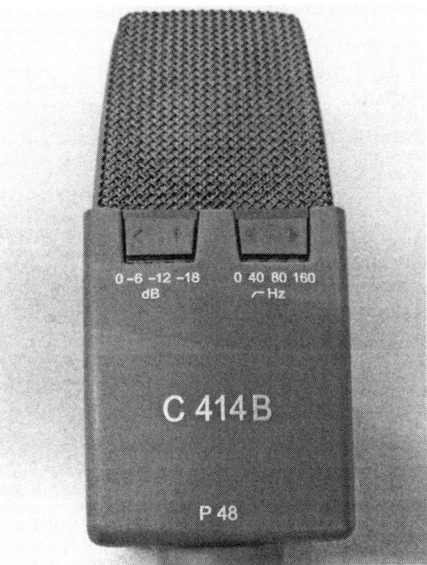

FIGURE 4.5

AKG C414 Bass Roll-off Select

Source: Courtesy of Tribeca Flashpoint Media AA; photo by the author

have a single, or possibly a two-position, selector, but here we have a wide range to choose from. Known as a *pad*, this switch is used to decrease the amount of signal level going to the preamp by a set amount and, on most microphones, will be in the –10 dB range. Some models allow up to a –20 dB pad. As you are setting up your session, if you are either hearing distortion from the mic or if you have to lower the gain on the microphone preamp to a major degree, the pad can be engaged to help keep things under control.

MICROPHONE EXPERIMENTS

Now that we have looked at the types of microphone design and how they pick up the sound (the polar patterns), what might be the best choice for your voiceover recording? Remember my answer from the beginning of this chapter? "It depends." Each situation is different, and the important thing to keep in mind is that each individual is different, as well. Just as a musical instrument can take on a distinct quality with one microphone and exhibit another quality entirely when recorded with a different mic, each person's voice will have a varying quality, depending on what microphone is used to record that voice. So how do we decide where to start?

A good way to learn about microphone characteristics and mic technique is to go into the studio and record yourself, and then listen back and analyze the results of your recordings. Also, this can be a wonderful ear training exercise and will help your critical listening skills. Try a variety of microphones, any that you can get your hands on, including those that you think would not be suitable for voice recording, and ask yourself: Which give the best results for voice recording? (And don't forget to also ask yourself, "Why?") Position yourself at various distances from the microphone—where does the proximity effect become pronounced? Move to the left and right of the microphone to experience and learn to recognize off-axis sounds. Record with and without a foam windscreen, as well as a pop filter. Place the microphone in various places around the studio space to learn where the optimal position for voice recording (the "sweet spot") is. Remember, the sound is going to change, dependent on mic placement within the room. Search out that "sweet spot" and remember it!

For these experiments, choose a short piece of copy, something around 30 seconds long. It could be anything: Advertising copy in a magazine ad, part of a newspaper story, a script pulled at random from a previous session, a poem, a page from a book. Remember, the content is not important, nor is your delivery (your voice acting skills). We're after the differences all of these variations make in the sound. Keep reading the same thing for each mic variation. This reduces the number of variables from one test recording to the next. And remember to mute the mic and turn off the phantom power when changing out condenser microphones! The differences from one setup or mic positioning to another may be subtle, but, with careful listening and practice, the change in sound and quality will soon become apparent.

I would recommend that you break the recording and listening sessions down into microphone types, distance from the mic, and placement in the room. This way you can easily keep your focus on each of these individual elements. Also, be sure to keep the signal chain consistent throughout your various experiments: Use the same microphone preamp for all of the mics you test and be certain to record everything at the same sample rate and bit depth. We don't want to begin introducing further variables into the equation—after all, we are testing microphones and acoustics, not mic preamps or the differences in audio quality between a recording at 44.1 kHz and one at 96 kHz.

Doing this type of experimentation by yourself is a good way to begin to learn what to listen for, and is a no-stress situation for you. You are not performing for anyone else and you can take your time (although you may put some miles on your shoes with all of the trips between studio and control room to start and stop recording). If you feel comfortable with it, have a friend assist you; it makes life easier, but is not necessary by any means. Just make sure to keep track of what you're doing and do each variation as a separate take, so that you can later identify what's what and what is most successful in your listening and analyzing session, once you're done recording. This is a very good method for learning your microphones and the studio space. You certainly don't have to do all of these variations in one go; however, the critical listening analysis should be done in one session period to give you the best idea of what is most effective with voice recording.

One thing to keep in mind, which I've mentioned before, is that each human voice is a unique instrument, so that a mic that works best for one person might be found lacking for another. A number of years ago, I was employed as a staff engineer at a studio that did a lot of voiceover work. A well-known voice actor by the name of Ray rented office space at this studio, so we had quite a bit of time to discuss some of our mutual interests, such as films of the 1940s and old-time radio. The point here is that I had a lot of time to just listen to the sound of his voice, outside a studio environment. Whenever he stepped into my studio to record a voiceover, however, I always found something lacking in his voice quality as it came through the microphone. Then, as now, my "go-to" mic for voice work was the Neumann U87. This microphone works well in a wide variety of situations, but for Ray it didn't capture the "Ray-ness" that I wanted, and so, for every session, I would put up the U87 and another mic as well. Nothing was giving me what I was looking for, and, after a time, I had gone through virtually every mic the studio owned. Finally, there was only one left, a ratty-looking old mic, no longer in production, and in such bad condition that the foam windscreen that completely surrounds the mic had disintegrated into dust. But, with the attitude of, "Nothing else has worked, what do I have to lose?", I arranged this orphan in front of Ray one day. Well, you know why I'm telling you this. It was gold! This dynamic microphone, with little output gain and a noticeable noise floor, was that magic sound I had been looking for on his voice for so long. To make a long story short, I sent this mic and the others of that model that we had back to the manufacturer for refurbishing, and they came back looking brand new. Unfortunately, owing to the design of this particular

FIGURE 4.6

Shure SM5B

Source: Courtesy of Shure, Inc.

mic, nothing could be done to raise the output gain or lessen the noise floor, but it didn't matter much if I could find the right applications for when to use this model. As it turns out, Ray bought two of them (offsetting the cost of refurbishing), and the studio had two as-new mics to put back into service. The moral of the story: If you are looking for that indescribable "something," keep looking. You just might find it in the unlikeliest of places. In this case, a beat-up Shure SM5B.

One thing that I can't recommend highly enough on this matter is the use of ear-training materials. It's certainly true that some people have a more acute sense of hearing than others, but anyone working in the studio can benefit from ear training. After all, you won't be able to identify problems if you don't know what they sound like. There are a large number of web sites and CD sets available—do a search for "ear training" and you are sure to find plenty to choose from. I have included a couple of web addresses in the Resources section at the back of the book. CD sets have a price, of course, but many web sites are free or charge a minimal amount for their use. To reiterate: You can't identify and fix problems, if you don't know what it is you're listening for to begin with.

With so many microphones available on the market, what are some of the most common that we find being used for voiceover? It's true that any microphone can capture speech and reproduce it, but remember that our goal is to record a full, robust, and *natural* sound. There are exceptions, of course, but this is true for 95 percent or more of all spoken-word recordings. As part of my research when beginning this book, I sent a short questionnaire to a large number of recording engineers and voiceover performers, asking what their preferred microphone was for this type of recording. The answers were what I had expected: The most common mic that was mentioned was the Neumann U87, by a wide margin. Although many other mics

FIGURE 4.7

Neumann U87

Source: Courtesy of Tribeca Flashpoint Media AA; photo by the author

were mentioned, the U87 was far and away the consensus winner for an all-around, "go-to" mic for most any situation in voiceover.

We'll get back to the Neumann U87 shortly, but let's take a look at some other common choices for voiceover recording, broken down by design type. What I am giving here is not a comprehensive listing, by any means, but these are the microphones that you are most likely to encounter and are commonly available.

DYNAMIC MICROPHONES

As I mentioned a bit earlier, dynamic mics are the most cost-effective design (in most instances), but aren't as transparent sounding as condenser mics. However, they may be a good choice on certain voices (such as my friend Ray) and in certain situations. To start, there is the Shure SM57 and its sibling, the SM58. These mics are very common in live-performance situations and can often be seen micing a drum set or used as vocal mics.

They are low cost and rugged and perform well in live situations, which is why many live sound engineers keep many of them on hand. For voice work, you may find that the SM57 works well on males, and the SM58 favors the female voice.

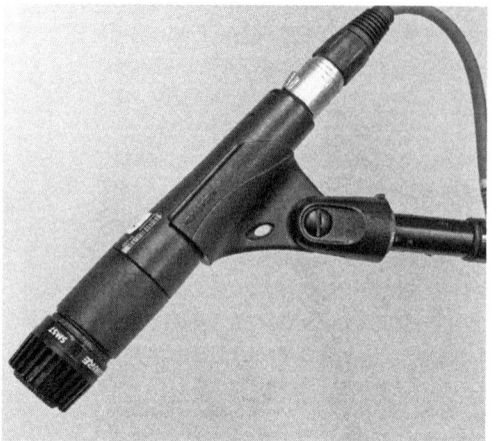

FIGURE 4.8

Shure SM57

Source: Courtesy of Tribeca Flashpoint Media AA; photo by the author

FIGURE 4.9

Shure SM58

Source: Courtesy of Tribeca Flashpoint Media AA; photo by the author

FIGURE 4.10

Shure SM7B

Source: Courtesy of Shure, Inc.

FIGURE 4.11

EV RE20

Source: Courtesy of Chicago Recording Company; photo by the author

These mics were designed with the stage performer in mind and exhibit a reduced bass response unless worked at a close distance. This design also helps to control feedback on stage. You may have to position your voice talent closer to the mic than if you were using a condenser mic, but they can be effective, especially if you are on a tight budget.

Two dynamic mics that are widely used in broadcast situations are the Shure SM7B and the Electrovoice RE20. Both of these mics have a large diaphragm and exhibit a wide frequency response. They can be a wonderful alternative to the more expensive condensers and are worth trying.

CONDENSER MICROPHONES

Moving up in performance from dynamic microphones, we next come to condenser mics. As explained above, condensers have a more open, transparent sound owing to the design of the diaphragm element and how it responds to acoustic energy. The transient response is more accurate, and the frequency response is generally wider than with a dynamic mic. For these reasons, a condenser microphone works very well in a wide range of voice applications. When you begin to research the condensers on the market, you will find a very wide spread of body designs and price points. Keep in mind that the mic that works great on one person may not be ideal for another, so that you may have to do quite a bit of testing to locate that one microphone that works especially well for you. I'll highlight a few of the better-known mics that are quite commonly found in a voiceover session.

As I've mentioned, I sent a questionnaire to a large number of recording engineers working in the voiceover field, and one of the questions on it was, "What is the one microphone that you would set up if you only had one mic to choose from?" By a very large percentage, the one mic that engineers use most often for voice work is the U87 from the German manufacturer Neumann. Nearly every studio has this mic in its collection, and the performance of the mic is hard to beat.

This ubiquitous microphone has a clean, clear, open sound and has the added benefit of having a very wide dynamic range. It has switchable polar patterns and a 10 dB pad, making it an ideal choice for many applications. The sound of this mic can be molded very easily with just a small amount of equalization, thus making it a good fit for any number of voices and actors. If I had to choose only one mic to purchase with which to record voiceovers, this would be it. It's my personal favorite for voice work, it's rugged, and it looks great. The professional engineers' choice, hands down.

If you're lucky enough to have nearly unlimited funds, this is what many consider the finest microphone ever made: The Neumann U47.

Take a look at a few old photographs of Frank Sinatra, the early Beatles, or Billy Holiday in the studio, and there is the U47, set up and ready to go. Its warmth and clarity make it legendary for vocal (and voiceover) use; however—if you can locate one—we're almost getting into the price of buying a new car with the cost of this

FIGURE 4.12

Neumannn U87

Source: Courtesy of Resolution Digital Studios; photo by the author

FIGURE 4.13

Neumannn U47

Source: Courtesy of Chicago Recording Company; photo by the author

mic. They are tube powered and are as rare as they are desired. Based on its predecessor, the U47 from Telefunken Electroakustik, this was the first mic with a switchable polar pattern and was based on the design of Georg Neumann from 1928. A true classic. (The latest listing for a Telefunken U47 that I saw had a price of $16,000.) Also from Neumann and popular with engineers recording voiceover is the TLM103.

In the price range of the U87, there is the AKGC414, also from Germany. Again, a very popular choice with engineers, this is the mic that I used for illustrations of switchable polar patterns and pads earlier in this chapter. Although many engineers like this mic for voice work, I've always found it to be a bit on the bright side for my taste and tend to avoid it. That is, unless the producer tells me that we'll be recording a television commercial and it should really cut through; then, this is my first choice.

There are a large number of newer models on the market these days, many from China, Russia, or Eastern Europe. These mics can offer excellent value for the price and should be considered in your evaluations when making a purchase. One that is gaining popularity is the Marshall MXL V88 (which was designed especially for

FIGURE 4.14

Neumannn TLM103

Source: Courtesy of Noise Floor Ltd.; photo by the author

FIGURE 4.15

AKG C414

Source: Courtesy of Tribeca Flashpoint Media AA; photo by the author

voiceover recording). Marshall also gives us the MXL M3B. If you are recording a variety of voice actors, you might also consider the MXL 9090; this mic is unique in that it contains two separate diaphragms, one of which is warmer sounding, the other brighter.

I've had good results with some of the mics from the Audio Technica line. The 4040 and 4050 are not only very responsive to the human voice, they are priced at a nice midpoint and should be seriously considered for voiceover applications.

Back in the US, Shure has many choices suitable for voice work in its condenser lineup. One popular model is the Shure KSM32, which you are likely to find in many mic closets today. Also widely used are mics from Blue Microphones, especially the Blue Mouse. Most microphones follow pretty much the same basic body design, and they tend to look very similar as a result. Blue Microphones has created an entirely new design statement with its line: All of Blue's mics are absolutely beautiful and quite unlike any other mic on the market. Blue seems to have a sense of humor about naming the mics, which I like, and the names aren't the usual number designations, but imaginative take-offs on the look of the product: the Dragonfly, the Mouse, the Yeti, Bottle, Bluebird, Bottle Rocket, Snowball, and so on.

FIGURE 4.16

Audio Technica 4040

Source: Courtesy of Resolution Digital Studios; photo by Bryen Hensley

FIGURE 4.17

Blue Snowball

Source: Courtesy of Blue Microphones

Before we leave the subject of condenser microphones, there is one other that deserves mention when we are talking about voiceover recording, and it's one that you might not initially consider. The Sennheiser 416 is a short shotgun mic and has become very popular with voiceover artists and engineers. Now, choosing a shotgun mic to record voiceover might sound like a choice loaded with problems: Shotguns are highly directional, and the smallest amount of movement of the head from side to side is immediately apparent. However, this mic has a powerful, "in your face" sound that is hard to beat for commercial applications. A word of warning here: Make sure that your actor has very good mic technique and is aware of the need to remain stationary when recording using this mic (due to the directionality of a shotgun). If you want impact with the recording, try this one out.

(For more about shotgun microphones, see Chapter 12: Recording Interviews and Roundtable Discussions.)

I'm not going to attempt a listing of all the condenser microphones that people use to record voiceovers; the list could probably run to a hundred different mics or more. Once again, my advice is to do your homework, ask plenty of questions, and try as many mics as you can get your hands on.

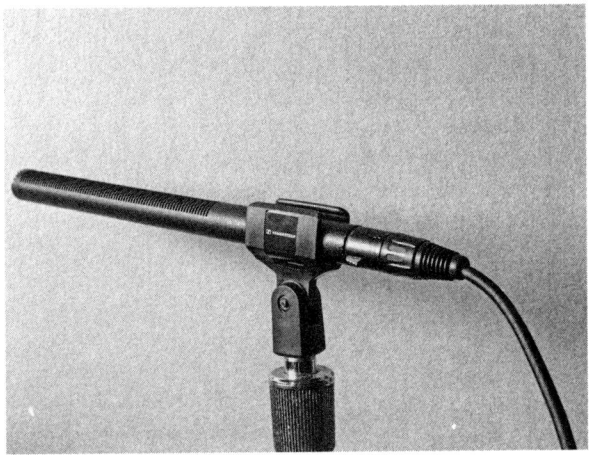

FIGURE 4.18

Blue Mouse

Source: Courtesy of Resolution Digital Studios;
photo by Bryen Hensley

FIGURE 4.19

Sennheiser MKH416

Source: Courtesy of Resolution Digital Studios; photo by Bryen
Hensley

I will confess that I haven't had much experience recording voiceovers with a ribbon mic. They were the standard for many years, until condenser technology began to evolve following World War II. Most of the classic mics we think of today come to us from the 1950s and are hard to top. Ribbons have made a comeback in recent years (especially in the music industry), but, because of my lack of experience with them, I am leaving them out of this survey.

DIGITAL MICROPHONES

In the past few years, a new type of microphone design has come onto the market— the digital microphone. We work in a completely integrated digital environment from start to finish in our productions, with the exception of capturing the sound. This is still accomplished using microphone technology that was developed over the past ninety years and that hasn't changed fundamentally in that time. Is it possible to replace the microphone as we know it with one that captures the sound digitally, thus completing our fully digital production path?

Let's review some basic facts about the world we live in. We, as human beings, are not digital creatures. We do not perceive the world around us in discrete quantized steps, but rather as an uninterrupted flow of sensations coming into our eyes, ears, and nerve impulses. In other words, we can say that we live in an *analog* world and perceive it as such. The sound we hear is created by air molecules vibrating and moving and tickling our ears. These air-pressure variations are then changed into electrical impulses by our ear mechanism and sent on to our brains, where they are perceived as "sound." A microphone works in much the same manner—moving air hits the microphone diaphragm, and that in turn creates a small electrical charge that we can now send along the mic cable and on to our recording device. The microphone is a *transducer*, changing one type of energy (acoustic) into a different type of energy (electrical). At the other end of the chain, we once again find a transducer, in this case the loudspeaker, which takes the electrical signal and alters it back into acoustic energy that vibrates the air molecules and, thus, passes on to our ears. So, to claim that we now have an "entirely digital signal flow" isn't really accurate at all. This change from acoustic to electrical energy is the fundamental basis of microphone design and construction (at least as we understand microphones today). So, is it even possible to create a totally digital microphone, and what do we mean by that term? There are a number of mics available today that advertise themselves as "digital mics," including the USB mics on the market. Let's take a quick look at what they're talking about when using that terminology.

Currently, the term refers to a microphone's *output* being a digital stream (we'll examine this output stream in further detail in a moment). The microphone element still has to change the acoustical energy of the sound wave into electrical energy before the signal can be converted into the digital realm, and this step remains the same as it's always been. In other words, the "front end" of the sound capture remains in the analog domain. The new digital mic design eliminates the need for an analog-to-digital (A/D) converter separate from the mic itself: Housed within the microphone body, an A/D converter is placed in close proximity to the diaphragm and allows for the digital stream to be sent directly out of the microphone. This has a couple of benefits: First, as mentioned above, it eliminates the need for a separate A/D converter. The manufacturer can design the converter that best matches the desired sound of the microphone, eliminating the variable of the user choosing a converter and coloring the sound of the mic. Because of this, the sound of the mic remains constant in all uses and better matches the manufacturer's intended sound, in much the same way that an active loudspeaker contains its own amplifier that the manufacturer deems best for the loudspeaker design.

Also, the mic preamp is housed in the body of the microphone. Once again, this element can be designed to match the sound the manufacturer intends. The benefit here, of course, is that the frequency response and overall sound of the microphone remain the same, regardless of where you might use it, eliminating the coloration of different mic preamps. Second, because the microphone outputs a digital stream, the problem of longer cable runs is now less of a concern. As we know, with an analog signal, there is a loss over distance through a cable (you might remember our discussion of the beginnings of digital audio and the developments pioneered by Bell Labs and AT&T that were focused on overcoming this very issue with a telephone signal). By streaming 0s and 1s, we can receive a cleaner signal at a greater distance.

Digital microphones output a signal known as AES42. This is not the same as standard AES digital, which is widely used in the industry, and, to connect the microphone to most digital devices, a converter must be used to change the AES42 signal to AES/EBU digital standard. There are a limited number of direct AES42 devices available now—the Sound Devices 788t comes to mind—and the selection is growing rapidly; if you want to connect a digital mic to your current gear, check to make sure that your equipment is AES42 compliant. Further complicating the picture, digital mics operate either as a Mode 1 device or as Mode 2. Mode 2 has definite advantages, although the mics are more expensive. Using Mode 2, the cable runs can be much longer, and this mode also allows for remote control of microphone parameters such as polar pattern selection, onboard DSP processing and sample rate selection. The digital microphone comes with a special cable (not the standard three-pin XLR microphone cable) and a converter (AES42 to AES/EBU). Microphone parameters are controlled via software.

Although currently quite expensive, digital microphones have many distinct advantages over their analog cousins, and the selection is bound to grow over the currently available mics. In my estimation, they are well worth checking up on.

USB MICROPHONES

With the advent of laptop computers and the digital audio software revolution in the twenty-first century, a new type of microphone has appeared. The USB microphone takes advantage of the fact that, these days, virtually all the recording we do is done on a computer. A USB mic has, instead of the usual three-prong XLR cable coming from it, a cable terminating in a USB plug and connects directly into the USB slot of the computer.

There is no need for mic preamps or A/D converters. This presents both a blessing and a curse, however. In principle, it sounds like the ideal solution. But, and this is a very important "but," USB mics rely on the mic preamp and converter that are on the computer's sound card, and, quite frankly, they're terrible. Avoid them if you want to produce professional audio. There's a very good reason for this, and it comes down to price. If you buy a computer for, say, $1,000, how expensive can any given component in the computer be? And there is a lot going on inside a computer, in terms of both hardware and software. If you were to purchase a moderately priced microphone preamp ($250) and a moderately priced A/D converter ($350) and put components of that level inside the computer, the machine would now cost $1,600—and the manufacturer just isn't going to do that if it hopes to

FIGURE 4.20

Blue Snowball

Source: Courtesy Tribeca Flashpoint Media AA; photo by the author

continue selling these machines. Add to that the fact that the inside of a computer is a very busy place, with a lot of electronic noise being generated. For these reasons, avoid using the preamp and converter that are on your computer's sound card.

Most USB mics have these components housed inside the mic body, but, once again, the components simply aren't up to the level you want to attain in a professional setting. Another thing to keep in mind is that the length of the cable is critical. It can't be too long (more than about 4 or 5 feet), or the sound quality is greatly compromised. In practice, this means that the mic is always in the sound field of the computer's fan and hard drive, and this introduces an element of noise into the recording that you do not want. Although fine for hobbyists and podcasting (for which they were originally developed), and coming in at a very desirable price, at this time USB microphones are not serious professional instruments, and they should be avoided for voiceover recording. As the saying goes, these mics aren't ready for prime time and should be avoided for serious voice work. Likewise, forget trying to get satisfactory results from the microphone in your laptop, tablet, or digital camera. It may be fine for recording Aunt Martha's birthday party, but it does not make the grade for full-bandwidth broadcast applications.

So, to return to our original question: "What microphone should I use to record a voiceover?" As you can see from this chapter, the choices available to you vary considerably. My advice is to experiment with a wide variety of microphones to find the one that works best on an individual voice.

The Engineer

5

THE ROLE OF THE ENGINEER IN VOICE RECORDING

Of all the subjects covered in this book, the one most difficult to understand is that of the role of the engineer. In this chapter, I will give you my ideas on what makes a good voice engineer and, hopefully, provide you with some tools that will help to make you an engineer that both producers and voice actors want to work with. To help us understand what qualities these folks look for and admire in a good engineer, I've enlisted the input from both voice talent and established producers who have worked in the industry for a long period of time. With the help of these experienced people, you may come to understand your role in the studio more fully, and their advice will give you some ideas on what to work on to make you desirable in engineering their projects.

In preparation for writing this book, I sent out a questionnaire to a great number of both actors and producers and asked for their input based on their many years of working in the field. These people, from across North America and Great Britain, were kind enough to offer their thoughts on the subject of the engineer, and I would like to take this opportunity to thank all who responded, even if their words didn't make it into the finished volume. All of their responses were valuable and informed me a great deal when I was thinking about this chapter. I think that hearing from these seasoned pros will help guide your thinking about the task of voice recording, and that hearing from them is a help in your understanding of what both clients and talent look for and expect from the engineer. A number of issues that they touch on will be covered in further detail later in the chapter.

For those of you who have your own project studio and act in the roles of both voice actor and engineer, much of this chapter won't apply to you. But, before you flip to the next chapter, I would recommend doing a quick read of this one; many of the same issues of dealing with the client and giving your best apply to all of us, regardless if we are doing it all ourselves or if we are an engineer in a commercial studio.

For the uninitiated, doing a voiceover session appears to be the simplest thing in the world. After all, you are generally working with one microphone, one talent, and one input on the console or controller. What could possibly be difficult about that? Understanding basic recording skills, acoustics, microphone operation, and so on is, of course, essential to performing your job. People expect the engineer to be technically savvy and to have the operation of the equipment down cold. But, beyond these skills lies another level of concern for the voiceover engineer. One microphone, one actor, and one channel on the mixer. What could possibly be more simple and straightforward? This is how the majority of spoken-word recording is done, and, on the surface, it appears there's nothing to it. Certainly nothing compared with a high-powered music session, right? Well, if there's one thing that I've learned in more than thirty years in the studio, it's that this can often be the most demanding recording situation that you'll be thrown into. Think about it for a minute: As I said in the Introduction, if you are performing with a symphony orchestra, you are playing with eighty or a hundred other musicians. But now step onto a stage with only your instrument and with an audience in front of you. There's no place to hide, and every moment is magnified. It's the same with recording the human voice. One microphone, one voice, and only one channel on the mixer. It's now just you and your skill to make or break this moment. Much of the information that we share as human beings comes through spoken words, and the majority of media that we come into contact with every day has a spoken component. Radio, television, movies, video games, even the annoying "Your call is very important to us . . . a representative will be with you shortly"—they all rely heavily on listening to someone pass information to us through words. And the important point to remember is this—some audio professional somewhere was responsible for recording all of this material.

COLLABORATION—COMMUNICATION—RESPONSIBILITY

Working in media of any kind is an exercise in collaboration. We can't function solely by ourselves in this business; others are necessarily involved, and, if you don't work well in a group setting, you won't make it in the recording industry. You will be working with artists, producers, directors, writers, and a host of others, all of whom have their own concerns and agendas. Learning to navigate your way through this maze of human ambition, insecurity, and ego is the perhaps the hardest part of becoming a successful engineer, but one that all of us have successfully managed to conquer. Learning how to read people and to get to the point of what it is that they're really asking you to do is a skill that takes time and patience to develop, and you will make mistakes and sometimes get the wrong message, but, over time and with experience, you will get better at this. It's just the same as learning your technology: The learning curve stretches out in front of you, but, before you know it, you are well on the way to performing your job without thinking about each individual step, and your skill level skyrockets, at times surprising even yourself. Noted sound designer and film editor Walter Murch makes the observation that the average Hollywood feature film takes approximately 400 *man-years* to complete. If you don't believe in collaboration and want to make the film yourself and have it released tomorrow, this means that you would have had to start on it before the Pilgrims landed at Plymouth Rock!

The key to successful collaboration is communication. Because of the number of people involved in any given project, information sometimes becomes lost or garbled, and the project suffers. The only way to arrive at a successful outcome is to make sure that everyone is communicating with all the other members of the production team, at all times. Phone calls, emails, talking, and social media all play a key part in the communication process. When you are scheduled for a session, immediately contact the person in charge (usually the producer or director, or their assistant) and begin gathering information that you'll need to make the project a success. Exchange contact information with everyone involved. The longer the project, the more critical communication becomes. For a 30-second commercial spot, the necessary information can usually be exchanged face to face, once everyone is at the studio. But, for a feature-length film project that might last for six months or more, the opportunity for face-to-face discussions becomes harder to come by. Always keep everyone involved informed about where you are in the process and what they can expect from you in the near future. Keep the questions coming and make sure you get the answers you need. *Assume nothing!* The more buttoned-down you can be prior to the session, and as the project progresses, the more successful the outcome and the more your reputation will grow. Although you may appear to be a pest to some, your attention to detail will only help the project in the end, and this whole business exists for only one purpose—to serve the project.

The better the communication flow, the less stress and sense of panic for you. I've always put communication high on my priority list in my work, and I think that it pays off in my feelings for my job. I've never viewed engineering as work; I really do think that what I am privileged enough to do every day is play with my friends. We work with really cool gear, work with creative and intelligent people, and make a creative and entertaining product. Where is the work in that? I truly believe that I'm one of the luckiest people in the world, because I don't work for a living. And a large part of why I feel this way is because I make sure that everyone is honestly communicating all the time.

Communication is part of your responsibility as an engineer. But responsibility goes beyond that. It addresses your professionalism as well as your skill. *Being* responsible also means *owning* your work and taking responsibility for all of your actions, good or bad. Stand up and own the choices you make, for better or worse. If you make a mistake, admit it, apologize, fix it, and move on. In Chicago, where I live, there can be no better praise for a person than to have someone say, "He's a stand-up guy." Being honest, ethical, and taking responsibility for all of his actions. He has *integrity*. He stands for something and isn't an opportunist who is out for anything he can get at this moment. Become that "stand-up guy." This concept scares a lot of people, and those are the ones who make excuses and constantly blame someone else or the equipment when things go wrong. In my work as studio manager, and now with the students in my classroom, I tell them that I am not interested in excuses: If you make a mistake, fix it. It's a waste of time to spend a 10-minute conversation telling me how or why it happened. You know it's wrong, and I know it's wrong, so fix it, and let's keep moving ahead and getting better. Those who don't take the responsibility for their work and their life are afraid of the possible consequences ("Oh, everyone will know that this is *my* fault!"). Don't fear responsibility; accept it. The perception that others have of you will only grow as a result.

This gets down to the subject of commitment, to the project and to the people that you're working with. Remember that what goes out the door at the end of the day has your name on it, and, as such, why wouldn't you want it to be the best that it could be? Collaboration, communication, and responsibility. These things show that you are committed to the project, and that you are taking ownership of the project. Now, I don't mean that you are taking over the project. I mean that you are fully committed to making it the best that it can possibly be. Treat each and every session as if it's the most important session you have ever done. And here's a little secret: *It is*!

MULTITASKING

I've noticed that those engineers who consistently produce great-sounding voice recordings have a number of things in common. First and foremost, they are world-class multitaskers. When doing any type of recording, we must be aware of many stimuli coming at us at once. With voice recording, we oftentimes must keep our ears open for problems and mistakes in the recording process and, at the same time, keep our eyes and concentration on a script and be on the lookout for missed or mispronounced words. Also, we must be aware of the narrative flow and make sure that the tempo and style remain consistent. At the same time, we are documenting the recording and keeping track of timings and good takes and being aware of possible editing choices we will make later. In Chapter 7, The Session, I'll cover many of these issues and how to deal with them in greater detail, but the importance of concentration and multitasking cannot be stressed too greatly: If you have one eye on your cell phone or tablet looking for that text from your friend about the concert tonight, you are not paying attention to the session, and, trust me, the client *will* notice. When you are in session, all of your energies should be focused on that session and nothing else.

Multitasking is essential during the session and comes with practice and experience. However, it doesn't appear out of nowhere; you should keep it in mind throughout all of your sessions. The number of things that you keep running in your head are staggering when you start listing them. First, of course, is the technical: Are my levels clipping? Is the mic properly placed? Do I hear any popped Ps or excessive sibilance? Is my workstation session set up properly? And on and on—it goes without saying that the technical aspects are your primary responsibility in the studio.

At the same time, you must pay attention to the *content* of what you're hearing. Are there any grammatical errors? Redundancies? Changes of tense? Do the words flow naturally for the ear? There are many writers who write for the eye and not the ear. What flows wonderfully on the page oftentimes sounds stilted or unnatural when read aloud (what I sometimes refer to as "brochurese"). Content matters; this is, after all, the information being passed to the listener.

Also, we must be aware of the editing process. Can we intercut two or more takes together? If you are recording a character voice or accent, is that voice staying consistent, or is the actor straying from where he or she started? Did the talent take an unnaturally long pause between sentences that must be tightened up? And, of course, you are keeping accurate track of take numbers and timings.

Is the talent becoming fatigued? Would a short break help her or the session vibe? Is everyone involved with the session comfortable, and do they have fresh coffee or water? Is the lighting comfortable for the talent, so that he can see the script but not feel like he's being interrogated in a police line-up? All this and more comes under your responsibility and is what sets one voiceover engineer apart from all the rest.

Yet another part of the job is acting as amateur psychologist to keep the session running smoothly and tempers on an even keel. Being keenly aware of the others in the room and being able to anticipate their needs is a big reason why these people come to you. And we have to keep an eye on the clock to ensure that the session is accomplished in the time allotted, and we don't run over and cost the client extra money. All of this adds up to an incredible level of attention to detail that separates the extraordinary engineer from the average, and keeping all of this in mind is how you keep working on a steady basis. So, in a sense, it's as much (or more) about these multitasking skills as it is about your recording skills. Of course, technical knowledge is the bottom line in any job in professional audio, but these other skills are what will set you apart from the competition.

If there is one piece of advice that I can give to someone doing any type of recording, it would be this: Always be prepared and ready to roll at all times. Digital audio is cheap—if someone is warming up or practicing, why not drop into record? You never know—sometimes that's when that magic take happens, and you don't want to miss it. At the least, you can delete those takes if you want to, and you're not out anything; but miss that one special performance, and it's gone forever. What this means is that, if the client says she wants the session to begin at 10:00 A.M., you have everything set up, the coffee made, the script copied, and a session built in your workstation, and you have verified that the signal path is working. At 10:00 A.M., you're all set to press the record button, capture every moment of magic, and come out with a superior product.

Being on time is of the utmost importance, as you can imagine. In any job you may have, if you are constantly showing up late, you won't have that job for very long. Working in a studio is certainly no different. In working on commercial productions, often the airtime is already purchased, before the producer and the voice actor walk into the studio, and the spot has to ship immediately. If you show up late, the chance of missing the air deadline is very possible, and, if that happens, you can kiss that particular client goodbye! So being on time is essential. And, as film director Woody Allen once observed, "80 percent of success is showing up."

If you do all of this, the client will return time and again; furthermore, the voice actor will appreciate your efficiency and hard work, and, the next time someone asks him or her for a recommendation on where to record the client's new project— well, who do you think that actor is going to suggest?

INSIGHT: Interview with Terry Schedeler, Writer/Producer, Dallas, TX

Q: What qualities do you feel a successful VO engineer should have?

A: In the work I've done over the years, in studios across the country, there are a number of key characteristics I've noticed that top engineers seem to have in common. (1.) By their actions, they make everyone (producer, talent, agency people, clients) feel as if theirs is the most important project in the world. (2.) They are incredibly laid back, yet laser focused. (3.) They adapt to the mood, tone, and pace of the session and anticipate the needs of the producer and talent seamlessly.

Q: What can the engineer do to make your experience in the studio go more smoothly and that would help you to capture the intent of your project?

A: Prepare. Listen. Know your stuff.

I think of a session as a performance in which the engineer plays a vital role as part of the entire ensemble. It all begins with preparation. Understanding the overall scope and goals of the session, creative approach, etc. Having the studio ready to go, even if you're backing up to another session. Knowing the players (producer, talent, client, etc.).

During the session, listening for things the writer/producer may not be listening for, especially in sessions where I've written the script. By nature, I'm going to hear what I wrote, which actually may not be the way it comes across to someone hearing it for the first time. An engineer who can ensure that the execution is technically correct, but also communicates the message as intended is a big asset.

The great engineers I've worked with know their stuff. By that I mean they know every acoustical characteristic of their studio, every quirk in their hardware, their entire music library and how to make their software and gadgets do amazing things. As I said, they are part of the ensemble. Their instrument is the entire studio, and their performance is the finished track.

Q: *Do you have any horror stories about what an engineer did wrong that ruined the studio experience for you?*

A: No horror stories. But I do remember one time when one of my all-time favorite engineers moved to the big city, and the new guy who replaced him insisted on telling anyone who would listen how much more he knew about recording than any of us who'd been doing this since before he was in kindergarten.

Q: *What is your feeling about the engineer/producer/talent collaboration during a session?*

A: As the producer, I'm always acutely aware that we're in the studio rapidly spending someone else's money on a project for which I am ultimately going to be held responsible. That said, the best work I've done has been a collaboration between the engineer, producer, and talent. These relationships are built on trust, experience, and judgment, even if these are established in the first session working together. We all earn our spurs, and the proof is there for everyone to hear.

Before every session, I have an idea in my head of what the project is going to sound like. The finished product never does. Something doesn't work the way I thought it would, so we adjust. Or the talent has a better way of reading it than I imagined. Or the engineer suggests a technique or approach I hadn't considered. So I know at some point, the way it's actually going to sound is the way it's going to sound, and I need to adapt to this reality—which is usually better than I had originally imagined.

My goal, as producer, is to create a comfortable environment in which everyone can do their best work, to set expectations and create a clear picture of what success looks like. And if I get some good advice along the way, to take it. My favorite engineers are the ones who follow my lead, understand that it is my responsibility to direct the talent and the session, yet who know how and when to steer me in a better direction.

Good sessions are like rhythm tracks. They find their own "groove." A good engineer is instrumental in making this happen, session after session.

Q: *Any other comments you would care to add?*

A: Don't be surprised if some of your most cherished friendships start in the studio. Talented people like to work with other talented people. That also explains why getting back in touch with them after many years is as if you'd seen them just yesterday.

INSIGHT: Charlie Pickard, Voice Actor, Columbus, OH

The greatest asset of a successful VO engineer is a pair of very discriminating ears! The ability to skillfully mic someone to his or her best advantage is absolutely important. It's up to the VO engineer to make him sound as good as is possible. Also important is getting balance between two readers or among three or more. Editing comes next, I think. The ability to blend portions of several takes into one great read is priceless. Today's "razor blade" is an electron wide, and incredible things can be done in the editing process. But it's not just cutting—it's *editing*: The matching, blending, and smoothing to make a take a winner.

A successful VO engineer is a consummate diplomat! Producers can be pains, readers can be prima donnas, and clients in the control room are the biggest pests since the rats left Hamelin! Diplomacy can be acquired, but it's tough. A good engineer wants to do his job and be done, but human beings are involved, and everything can go downhill fast. Like it or not, the VO engineer is the go-between for the talent and the producer, guiding the voice toward what the engineer hopes the producer wants, keeping the producer aware of what he or she is demanding of their talent, keeping them from expecting more than is reasonable. I've freelanced for thirty-five years, and my mantra has always been the same: There are only two people a voice talent needs to cultivate—the person in charge of Accounts Payable and the VO engineer, and not necessarily in that order. Always follow the rules for submitting an invoice, no matter how arcane. And know that the only true friend you can have at a studio is the engineer. How he handles your mic can make or break you; how he deals with the producer can make your job a joy or hell; your relationship with the engineer can relax you and help you do your best, or put you on edge and make you miserable.

I would encourage anyone who wants to be a VO engineer to be as friendly as possible in the circumstances. Try to put the voice talent at ease and advise him or her of the quirks of the producer, if the talent has never worked for that person before. Keeping things as loose as possible (within reason!) really helps the session turn out well. And being the trusted intermediary is a great skill.

Two quick cases where the engineer saved the day—the engineer was the same in both cases: In the first instance, I was recording a spot for a well-known department store, and the copy included the word "accessories." The producer, an otherwise great guy, insisted I pronounce the word "a-SESS-ories." It took the engineer 10 minutes to convince him otherwise, but he did, and saved the spot. The other instance of duty above and beyond came one afternoon. I was to cut a spot for a client I can't remember, but the producer was an old hand who knew his job and was slow to lose his cool. He was accompanied by a young man who was being broken in as a future producer. There I was in the studio, with the producer, the engineer, and the young man in the control room. Suddenly, in came the account exec and three people from the client! I could see the producer's face getting redder by the minute, but somehow the engineer, with his skill and professionalism, got us through the session without a murder occurring. I took my sweet time leaving the studio to let things clear up in the control room! A good engineer is a pearl of great price!

I've sort of touched on the engineer–talent relationship earlier. Let me just say that technical skill is the foundation, of course. But diplomacy and the ability to relax other people in a stressful situation are just as important. Being able to learn the idiosyncrasies of both the different actors and the various producers you work with is well worth the effort. And one other thing, of supreme importance: The VO engineer is the king of his or her domain. The producer/agency is paying money, and their wishes are important; the talent's abilities and feelings are important, too. But the smooth operation and meshing of the technical and the human is in the engineer's hands. And so is the continuing smooth operation of the control room. The engineer's word must be final!

One thing that is sure to elicit groans from all involved is for the client to show up for the session. These folks assume that, as they're paying for the session and its product, they have the right to come in and begin taking over the session. This is as logical as going to a fine restaurant and, because you are paying for the meal, walking into the kitchen and telling the chef how to prepare the food. Look, you hired us because we are creative, talented people who do this for a living and, most likely, have spent years learning our craft. The last thing we need is to have someone come in and start micromanaging every decision and every action that we take. The client may own the company, or manufacture a wonderful product, but they don't understand that every session has its own "groove," or flow, and that it usually takes a couple of read-throughs for the rhythm to be found. Nor do they understand that the powerful and nuanced finished product is quite often the result of a combination of pieces from numerous takes edited together, and that the engineer and the actor, not to mention the producer, are very well aware of which pieces will fit together in the completed edit. The assumption is that the actor can come up with the perfect read the first time through the script ("Why do we have to do it again?"). The client (most times) doesn't know how to direct an actor, nor has the technical vocabulary to communicate his ideas with the engineer or producer. They most often give confusing, conflicting direction to the actor, question every move an engineer makes, and second-guess the producer with every decision made. They simply can't get out of their own way and end up with an over-budget end product that isn't nearly as good as it could and should have been, if only they had remained in their office. And, of course, this state of affairs is all because those people doing the session didn't know what they were doing! In short, they are quite useless in the session. My auto mechanic has a wonderful sign hanging on the wall of his shop: "Labor—$45/hr . . . If You Watch—$90/hr." Oh, how I wish we could send out a similar notice to the clients thinking about coming to the session! I've often, and only half in jest, said that 80 percent of our job is saving people from themselves, and I do believe that firmly, but politely, telling the client to just sit back and watch is doing them (and everyone else) a world of good.

BUILDING A SENSE OF TRUST

Both Terry and Charlie touched on a very important point, and one that I cannot stress strongly enough: The key to a successful career, not just as a VO engineer but in anything we might do, is to establish a sense of trust with the people you work with. The old saying, "You're only as good as your last job," certainly is true. Now sure, we all have off days, and we hope that one of those is not the first time you work with a client, but, if the majority of the time you're on your game and you keep the session running smoothly, the client will want to come back time and again, because they know that you are going to give them your best. In a conversation I had with studio designer and engineer Dave Hampton (author of *The Business of Audio Engineering*), he relayed a very valuable piece of wisdom: "Trust ain't an app. It doesn't cost 99 cents, and you can't download it." Trust is earned through repeated effort, every minute of every day. It's hard to earn and is the most valuable commodity that you possess. Guard it with your life! If you say you're going to do something, then carry through and make sure you do it. Remember, at the end of the day, you have only one thing, and that is your good name.

And here's another thing to keep in mind that I learned from a studio owner that I worked for a long time ago: There is no such thing as an insignificant session. Every day, treat every session you do as if it's the most important thing that you've ever done. It's very easy to see an inexperienced and bumbling producer trying to get a handle on turning the product out and think, "This bum's not going to last till the end of the month—why should I care?" Easy enough to blow him off and move on. But, and here's the moral of this story: I once had a producer like this come in; it was evident that he didn't have a clue as to how to arrive at where he wanted to get to. Instead of just pushing the buttons and watching the clock, I gently took over the direction of the session, explaining each step as I was doing it, from recording through editing and mixing. Over the course of a couple of sessions, I gained his trust because of the interest that I took in his project, and we became close friends. He came in about every two weeks with another long session to accomplish, and, after an amazingly short period, he turned into one of the finest producers that I knew. After three years, this client had turned into a $7 million account for the studio! Not bad for a small, independent studio, and an account we would not have had if I had just hustled him out of the studio because the first session was a pain. I used the occasion to build his trust in both the studio and me; we kept our word on pricing and deadlines, were fair and open with him, and treated him as the professional he was (even if he wasn't aware of being one yet). Trust wins another one!

Always keep in mind that we are in the service business (and that this *is* a business, no matter how much we claim that we're in it for the art of it all). If someone treats me fairly and respects me as a person, and not just as a button-pusher or as an engineer, I will repay that with a respect for their work and a commitment to their project and their vision for the project. Being in a service industry, we work on what is called a "fee-for-service" basis. That is, I agree to provide you with a service (recording, editing, and mixing whatever audio you require) for an agreed-upon price. It makes no difference if you are an independent contractor or the owner of a huge multiroom complex; this is the business that we live in. As a practical result, your service to the client is paramount, if you want to stay in business. How you treat the client and live up to what you promise to do is the determining factor in your success.

PREPARATION

Preparation is key to a successful session. Before the session, try to get as much information as possible from the client about what the session will entail. For instance, how many actors will be recording at once? Is this session a radio spot or a piece of long-form narration? What types of deliverable are the client expecting? What file format is required for the deliverables (including sample rate and bit depth)? Are you going to be receiving any materials from another studio (such as a music track or a previously recorded actor that will be cut into the final track)? Discovering all of this and more will, first of all, make your job much easier and help the session run more smoothly, but, more importantly, it will tell the client that you care about them and their project and you are working overtime to make sure that you give them what they want. By asking as many questions as you can beforehand, you will have everything set up and ready to go the minute the client and voice talent walk in the door. This is crucial if you are recording radio or television commercials. Try to get the script beforehand and then, with stopwatch in hand, read it aloud. See if, reading at a comfortable pace, it fits into the allotted time (30 seconds, 15 seconds, what have you). I stress the importance of reading it out loud, because reading it silently or under your breath will lead to a false sense of timing; we read much faster with our eyes than we speak, and you must be sure that your timing is accurate. If you detect a problem, alert the client immediately. Calmly explain the situation ("This copy runs about 10 seconds too long—can you cut some words?"), and once again you show that you are thinking about them and only them. Everyone likes that feeling! After all, time is money, and, if you can catch something up front that saves them money in the studio, they'll really appreciate it. Mistakes in the script aren't your fault, and, if a make-up session is needed to correct things, make sure that you have stated up front that any re-dos will be charged over and above the price that you have agreed on for the session.

Before you can begin recording the spoken word, first you need a script. Now, this isn't your responsibility, but you should be aware of what the script represents. A well-written script is like the score to a musical work—it's a guide and it is open to interpretation by the artist. I don't necessarily mean changing the notes on the page (or, in our case, the words), but rather things such as the tempo and emphasis of the performance. The writer who is experienced in writing voiceover scripts knows that there is a large difference between writing for the ear versus writing for the eye. Things that look good and make sense on paper do not necessarily strike the ear as conversational and logical. This idea of writing for the eye versus the ear is a difficult thing for writers to learn, unless they regularly write for actors, e.g., writing screenplays and the like. If you are presented with a script written for the eye, there may not be a lot you can do about it other than (politely) point out that something would make more sense if worded in another manner. Be aware that there will be times that no changes can be made; remember that your client most times has a client that they are working for, and that the ultimate client may have approved the script with no changes to be made. Also, there may be some type of legal consideration that you are unaware of, and that is why the script is worded the way it is. (My current favorite is from political advertising, where the candidate *must* say, "My name is Joe Schmoe, and I approve this advertising." Why not "advertisement"? Some Board of Elections Commission lawyer somewhere, I suppose.)

When it's time to start the session, everyone must be working from the same version of the script. I've seen confusion during sessions because there were two or three different versions of the script floating around the control room; no one is then quite sure which is the correct version, and time is wasted trying to contact the writer or whoever may know the answer. It has always amazed me how many times we have recorded the wrong version of an "approved" script, only to have to book another session to get the correct words down. I ask myself, "Wasn't anybody paying attention?" Apparently not.

When discussing the script with the client, remind them of a couple of formatting suggestions that will greatly help the talent and the engineer: Have them double-space the lines of type and leave wide margins on the page for notes to be taken; avoid breaking sentences over a page break (an invitation to paper-shuffling noise); spell out any words that may be confusing (for instance, do we want to pronounce the client's name as "Inc." or "Incorporated"?); the same goes for numbers (105 could be read as "one hundred and five," "one hundred five," "one oh five," and so on—which is preferred?); and, if the script consists of multiple pages, have the page numbers clearly marked. All of these things will help the session run in a much more efficient manner and avoid excess billing charges for the client. Above all, make sure that the script is laid out in a clear and concise way that is easily read on the first run-through.

It should be added here that part of your preparation is staying current in the technology. After all, we work in a technological profession and we are expected to know the latest and greatest. Gather as much information as you can get your hands on about what's happening in the industry and what the hot new products are. A case in point: A long time ago, I read an article in *Mix* magazine that said that the technology was now available to be able to synchronize an audio tape recorder to a video machine using time code. I was curious about how this worked and did a bit of research (this was long before the Internet, and so the research was more difficult to do, but, by talking to a large number of people and reading as much as I could find, I started to get a handle on the process). After learning enough about it to be able to intelligently discuss the subject, I brought up to my studio owner that, if we invested in this technology, we could work to picture, recording directly to an audio tape machine and not have to go through the cumbersome and expensive process of recording to magnetic film. He gave me the go-ahead to contact the manufacturer of the synchronization equipment, who then came in and gave us a demo. The owner purchased the gear, I spent a couple of months really getting it down, and we were in business as the only studio in the area being able to work in this manner. Over time, I started to become known as the time code fix-it guy, and people would come to me if they were having problems with their synchronization. I then got a job at another studio precisely because I knew how all of this worked, and the owner was looking to branch out into working to picture. I spent eight years at that studio and was then hired by yet another studio because of this knowledge. So, for over twelve years, I kept my career thriving because of the initial curiosity I had over that long-ago article, and I kept up with advances in the technology. After working at the third studio for a number of years, I elected to begin a freelance business, and even then people would hire my services for their synchronization jobs and problems. All part of the preparation for the session—even if that session is twelve years in the future.

INSIGHT: Interview with Linda Wolfe, Producer, Chicago, IL

Q: What qualities do you feel that a successful VO engineer should have?

A: Obviously technical skills, but also the ability to "Zen" the process. If the process isn't second nature first and foremost, then all the creativity in the world won't matter. Collaboration and having ideas is what it's all about as well. A monkey can be taught to push buttons. Producers need solutions to challenges and problems that might arise, and they always do. An even temper, not being too talkative, an ability to continue working while all hell is breaking loose around you. An ability to stay "removed" from the chatter and idiotic banters of young creatives or clients who have silly ideas. The ability to make everyone feel like they are being heard and yet staying focused.

Q: What can the engineer do to make your experience in the studio go more smoothly and that would help you to capture the intent of your project?

A: Do your homework! Understand and study the project before it arrives at your door. Have everything loaded, review takes and tracks if possible. If you are recording VO, understand the intent of the spot. If music, have some conversations with your client beforehand if possible to discuss direction and challenges. *Be ready to go.* Although we all love being in a studio, time is money, and we hate wasting both. If you are doing sound design, have some ideas for sfx and stock music. Time the script beforehand if you are working on a spot. *Do not own* the session! Arrogance is not pretty. You will have great ideas that will not be accepted. *Be prepared to give up your darlings!* It's not life and death. Although you may disagree with a producer or client, in the end it's their dime. I love working with an engineer who can "discuss" why something is a good or bad idea with me. I like to be challenged, but in the end, it's *my* gut that will make the decision.

Q: Do you have any horror stories about what an engineer did wrong that ruined the studio experience for you?

A: *Yes!* An un-named engineer "lost" an entire recording session's takes. *Back up!* Additionally, I hate slow engineers who are unsure of what buttons to push. It drives me crazy!

Q: What is your feeling about the engineer–producer–talent collaboration during a session?

A: It's a partnership. First you have to have the "right" talent. If you have that, you are a third of the way there. If you have a seasoned producer who can get what is needed in less that a million takes and who understands the intent of the message, you are two-thirds there. If you have an engineer who is on the same page and can put it all together, and most importantly has the "timing" to get it right, you win. There is nothing worse than a good script that has bad timing: Pauses that are too long, bad edits, clipped voices. I hear everything. If you can hear something and it's not perfect, *fix it!* A good engineer is a trusted partner.

Q: *Other comments?*

A: When you start getting clients, you will have some that never come back. Don't worry. It's like dating or finding a good doctor. Only be concerned with the ones who think you walk on water. I also use different engineers depending on the content of the project: Comedy, long-format, tons of effects, lots of stock tracks. It would be great to be a master of it all, but most times you can't. Understand what you are good at first, nail that, and then move on. It takes years to be great. It took me about fifteen years before I *knew* I was one of the best producers in the city. Be patient. Practice, practice, practice. *Fast* and *good* is what it is all about.

THE TEAM

All three of these guests brought up an interesting point, and something that you should carefully think about. Life in the studio is an exercise in collaboration. I've worked with producers who wanted me to only push the buttons and never open my mouth, and I've worked with producers who considered me a co-producer on their project. In the end, I'm fine with whatever role is expected of me, but I firmly believe that everyone working as a team, with open and honest communication, is the way to achieve a superior product. If you are comfortable with your technical skills, they become almost second nature, and you won't spend excess energy thinking your way through a session. It takes time, patience, and practice to get there, but, once you've mastered your equipment, your work flow will be much faster and much more creative. Then, and only then, can you begin to be a collaborative part of the recording team and have the client and voice talent really start to trust your skills and your decision-making. For me, this collaboration is by far the most enjoyable part of engineering. If I hear something that I think can be improved, I don't hesitate to point it out. And Linda Wolfe brings up another point: In the end, this is the client's party and their dime, and their decision is final. If you don't agree, do what they ask for and move on to the next session. Many times I've sat in a studio with a young producer who thought he or she had all the answers and would ask for something in a particular way. I would patiently (and politely) try to explain that it wouldn't work. "For me, this is going to work—nobody else knew how to do it!" comes the response. Again, I would explain that I've had years of experience doing this type of project, and it's been tried before, and whatever they are asking for never works. "I'm telling you, for me this is going to be great." So, OK, hotshot, I'll give it my best go. And of course it doesn't work, and then what happens? They blame you! Get used to it; this scenario will play itself out countless times.

There will be situations where you can tell what the client's final goal is, and yet they have no idea how to get there, or you know that you have a better, more creative approach to something than they are directing you to do. In this case, I advise you to give the client what they want, even if they don't know it yet. By this, I don't mean to try and impose your own vision on the project, but rather to take it in the direction that you are certain is the correct way to go, and, with a bit of luck, the client will see where you're heading and agree that you may indeed have a better idea. But, as Linda pointed out, be prepared to kill your children! All of that great editing and sound design may be thrown out, and all you can do in that case is move on and give them what they want.

Part of the job of a studio engineer entails working with all different kinds of people. I've always thought that being a successful engineer is about 20 percent technical skills and 80 percent people skills. As I mentioned above, being an amateur psychologist (and sometime mind-reader) is a prerequisite. Not every minute of every session is going to be spent recording takes; there will be times of casual conversation and getting to know one another. Because your clients and talents are all different people, with their own interests and experiences, developing your own wide-ranging interests will help you relate to them and take an active part in conversation in the studio. Read. See more movies. Learn a bit about history or science. It's actually fun and enlivens your life outside the studio. For most of my life, I didn't pay much attention to professional football, but, every Monday through Wednesday, everyone in the studio was talking about last weekend's game, and I was left out of the conversation and felt like there was something wrong with me. So I started to take some time on Sunday afternoons and watch the home team and began to learn the players and the strengths of the various teams in the league. After a short while, joining in the conversation about the game became easy and fun and helped me build a relationship with a diverse group of people. All part of the job.

I always try to instruct an apprentice or beginning engineer on when to speak up and when to keep their mouth closed. As with anything else in the studio, this takes time and practice to master; diplomacy doesn't always come naturally. You have to gain the experience to be able to read a situation and react accordingly, and you will make mistakes. If so, apologize and move on. You have just learned something about the client or actor with whom you are working, so file it away for the next time you work with/for them. At the end of the day, your feelings are not the important issue—it's all about the project.

This brings me to my next point: Leave your ego at home. What with the clients and the actors, there will be more than enough ego to go around when it comes time to do the session, and yours is the least important. We don't need yet another prima donna in the studio, and constantly attempting to tell everyone how great you are is a sure way to drive them from your studio. Know that you're great, but be humble about it. If you do your job correctly and creatively, everyone will acknowledge your expertise and skill; you don't need to advertise it. Always keep in mind that this is *not* your project—you are there to help bring the vision to life as a part of the creative partnership.

A long time ago, my father told me that, if you don't like your job, go find something else to do. Life is too short for you to be miserable while you're trying to make a living. The recording profession in one of *passion*, and one that most of us feel that we simply *have* to do. This passion will carry over into everything you do and will be evident in your work. I've always said that I may not be the most technical of engineers, but I am certainly one of the most passionate. I learn everything I can about this industry—its history, new products, new (and old) ways of doing things. All of this shows in my work, I think, and makes me want to leap out of bed in the morning and get back into the studio. As far as I'm concerned, there is no better place to be than in a studio, and that's because I've developed a passion for doing this work. If you feel that you can't possibly do anything else except record and mix, you're on your way to being a fine engineer.

The Studio

6

Each type of session makes its own demands on the studio space and the setup of the studio. Music sessions rarely need video playback, whereas an ADR or Foley session absolutely needs picture in the studio space. A commercial recording session may or may not have such a need. When discussing setting up the room for voiceover recording, keep in mind that this type of session has its own needs and is unique. Every engineer will set up their studio in a way that works for him or her, and I wouldn't attempt to prescribe a one-size-fits-all solution, but I can offer some advice that applies to most situations and leave it up to you to customize your setup so it works best for you.

A FEW SIMPLE TRUTHS

No matter how you prefer your studio to be set up for a voice-recording session, there are some things that will not change, no matter how you set up the room. The physics of sound and the acoustics of the room remain a constant (see Chapter 3, Room Acoustics). The size of the room will certainly play a role in how your recording sounds and it cannot be changed (except by the addition of gobos and the like, or some other method of altering the shape and size of the space). How a microphone functions and how it captures and reproduces the sound remain constant, although microphone placement within the room can alter the sound and quality of the recording (see Chapter 4, Microphones).

Above all, the studio is a place for humans to work and create and must be treated as such: Comfort and ergonomics must be taken into account. Lighting, heating, air conditioning and so on are very important considerations and should never be taken lightly. After all, for the talent to give his or her best performance, they must first of all be comfortable and relaxed, and a room that is poorly lit or is much too warm and claustrophobic is not conducive to being a creative work space. Although all of this may seem perfectly obvious, I've witnessed engineers doing a voice session in a cavernous music studio, and, although it may have cool "mood" lighting that adds to the vibe of a music session, the lighting makes it difficult to read a script, and there is no video playback if needed. Furthermore, because the studio was large enough to accommodate a great number of musicians at the same time, having just one person standing alone in the room leads to an isolated feeling, and the talent doesn't feel connected to what is happening in the control room. A single individual standing by herself in the corner can be pretty lonely. One only has to take a look at the Beatles recording their early music, set up in a corner of Studio B at EMI Abbey Road studios, which is large enough to accommodate an entire symphony orchestra, to get a small feeling for what this experience must be like.

One last point: In those studios where I have been the lead engineer or studio manager, I always tell my assistants and the other engineers that no one notices a clean studio, but *everyone* notices a cluttered and dirty studio. So make sure you keep the studio space neat and tidy. Now, I'm a bit of a fanatic on this score, but everyone should be able to keep cables neatly coiled and be capable of clearing out the used coffee cups from two sessions ago.

So, in this chapter, we're going to take a look at some ideas for setting up the studio space for maximum talent comfort and creativity.

LIGHTING

The first step in setting up the studio space for the voiceover actor is making sure that the lighting level is appropriate for comfortable reading without eye strain or glare. There are two types of lighting you should be cognizant of: The overall room lighting and the lighting of the script itself. Let's address general room lighting first.

The light level in the room should be comfortable for most activities and not be too bright or too dim. I do appreciate having the room lights on a dimmer to set the mood of the script—provided the script light is adequate for illuminating the page. Many actors prefer having low-level ambient lighting when they're in the studio and feel that it helps the performance, especially if the read is to be intimate and quiet. Considering this ability to dim down the room lighting leads us immediately to what type of lighting we're working with. Fluorescent lights usually can't be dimmed and they introduce a great number of other problems into the space. First of all, by design, fluorescent lights flicker, and this can cause eyestrain in a great number of people. As well, they are quite bright, and the color balance of the lighting element is harsh. Although good for work lights when wiring a room, for example, they are not recommended for overall ambient lighting needs in the studio. Another thing to keep in mind is that fluorescent lights tend to buzz, owing to the ballast used to power the lights, and this buzzing sound is easily picked up by the microphone and is very difficult to remove from the recording. So, fluorescents are out.

Incandescent lights (regular light bulbs) are by far the most common to find in the studio. They are low cost, give good light balance, are dimmable, and are easily changed if one burns out—all good reasons to choose this type of lighting design. By altering the wattage of the bulbs, you can adjust the overall level of lighting in the space and then, with the addition of a dimmer on the light switch, you can raise or lower the room brightness to match the required level. One word of caution here, however: When installing dimmers on the lights, be very aware that dimmers often cause a 60 cycle hum or buzz in the audio lines, and so you may have to go with theatrical dimmers, which use a resistor to prevent this noise—a bit more money, but well worth it for a clean audio signal. A downside to incandescent lighting is that it tends to throw off quite a bit of heat. Put your hand over a light bulb in your home and you'll see what I mean. If you have a dozen or so light fixtures in the room, all throwing off that much heat, you can see how the temperature levels in the room rise pretty rapidly, and this will require a greater amount of air conditioning to be introduced into the room to counteract it. This, in turn, presents its own problems in the form of additional noise coming from the ventilation system, but, far and away, the incandescent lighting system is the prevalent solution to studio lighting.

However, things are changing—there are some new options that you should consider. Compact fluorescent light bulbs have some positives going for them: They fit right into the incandescent fixture, so you don't have to do any rewiring. They last up to six times the life span of an incandescent light and operate at a cooler temperature. However, they still have the flicker and color-balance issues that are inherent in fluorescent fixtures.

A new player on the scene in the past few years is the LED (light-emitting diode) light. These lights use an extremely small amount of energy and are surprisingly low in operating temperature (hence, less cooling is required), and the life span is incredible compared with the standard incandescent light bulb. You can put them on dimmers, and there is an almost infinite range of colors of light that you can have with them.

When John Storyk designed Electric Lady Studios for Jimi Hendrix in 1970, the artist wanted to have stage-type lighting in the studio. The solution was to paint all of the studio walls white and then hang a theatrical lighting grid in the studio, so that the walls could be washed with a wide variety of colors, as fitted the mood of the moment. Needless to say, they were very hot and used a large amount of energy to achieve the effect. Now, with the advent of LED lighting, we can achieve the same effect in a much more cost-effective and energy-efficient manner. Here is John Storyk on the subject of studio lighting:

INSIGHT: John Storyk, R.A., Principal—Walters–Storyk
　　　　　　Design Group

Let's go down the list here: Fluorescents are just ugly. These rooms are usually small, so, unless you use them as cove lights, stay away from them. And they do make noise, because the ballast is usually on the fixture. So, stay away from them, period. There are very, very cool incandescent lights, and most of the time if you put them on dimmers they're going to be dialed down. We're now starting to use a lot of LED lighting—everybody is. And they're terrific: They cut down on your air conditioning because you use a fifth of the energy, and they don't make any noise. And it's going to be the law pretty soon, so you might as well start using them. You can put them on dimmers. They come in color shades and they now have bulbs that are RGB bulbs—the bulbs themselves have all the colors. They're very cool, no two ways about it.

Dimmers work—just use the right dimmer and the right fixture. Our rule of thumb is: "Stay away from low voltage, just stay away from transformers." Get the transformers out. These new LED lights are really good. They don't make any noise.

The next issue to address after the ambient lighting in the room is the illumination of the script itself. For this purpose, it is best to have a small light mounted directly above the script that isn't affected by overall room lighting levels and colors. Many people use a clip-on reading lamp for this purpose, but I don't like them for two reasons: First, they create a "hot spot" on the script, and glare is produced from this small area, and, second, they are not as sturdy as we would like in a studio setting. For lighting the script, the best solution is a light from the Manhasset music-stand company (www.manhasset-specialty.com). The model 1000 music lamp provides consistent illumination across the entire page (thus eliminating any hot spots), is glare-free, and is shielded from the front. Thus, if you are viewing the actor from

FIGURE 6.1

Manhasset Model 1000 Music Stand Light

Source: Courtesy of Manhasset Specialty Company

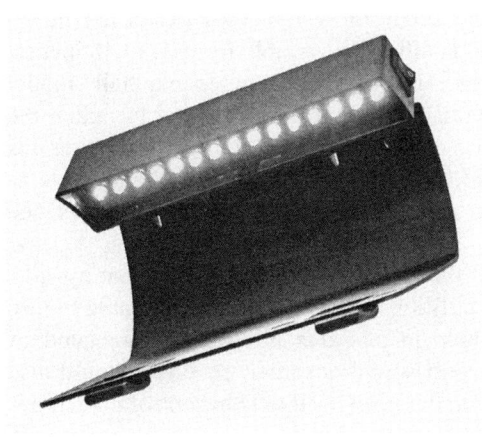

FIGURE 6.2

Manhasset Model 1050 LED Light

Source: Courtesy of Manhasset Specialty Company

FIGURE 6.3

Manhasset Model Light and Stand

Source: Courtesy of Manhasset Specialty Company

the control room through the window, you won't be bothered by looking into the light when addressing the actor. The lamp operates on a long, slender 40-W bulb and comes with an 8-foot-long AC cord. The light attaches to the music stand with two strong clips, making it very sturdy and wobble-free. This light was designed for orchestra and band musicians on stage and is ideal for our purposes in the studio. Keep in mind that your voice talent needs all you can provide in the way of help in order to deliver the best performance, and it all starts with being able to see the script clearly, without shadows or glare.

The Manhasset company also makes a new product that is a nice combination of lighting and energy efficiency: The model 1050 light. It has the shield/reflector of the older model 1000, but incorporates LED lighting powered by a rechargeable battery. The estimated battery life is eight hours, and this stand light provides a very comfortable lighting level and color balance for the talent.

THE SCRIPT EASEL

We don't want the actor to have to hold multiple pages of a script while trying to read and so rustle the paper, and so something must be provided to place the script on while he performs. The natural solution is a music stand, and music stands come in a few basic models. The first is the portable and collapsible wire music stand familiar to anyone who has played in their grade-school band. These stands are flimsy, and the maximum height they can be raised to doesn't work well with standing adults. Again, we turn to the Manhasset company for our answer. The model 48 Symphony Stand answers all of our basic needs for a voiceover performer in the studio. Sturdy and good-looking, the stand is infinitely variable from 26 to 48 inches, for a maximum overall height of 60 inches. This stand also comes in a "tall" model that extends to 60 inches (72 inches overall height). It is well suited to match the music stand light mentioned above. The easel itself tilts from 0 to 180°, making this stand an ideal candidate, not only to hold the script, but to tilt to 90° and serve as a lightweight table to hold bottled water, pencils, jewelry, and so on, when placed directly next to the voice talent.

The easel should accommodate two to three full pages of the script side by side, to help the talent avoid excessive paper-shuffling noise. A good talent is able to turn pages silently, but giving them a little help in this area is always appreciated. A piece of paper doesn't make a very efficient sound absorber, so there should also be some sort of absorbing material placed on the easel itself to help control reflections. This can be a piece of carpet remnant cut to fit the size of the easel or, better still, a piece of foam. The padding on the stand also helps to mitigate the sound of paper rustling when pages are being turned. Although necessary, this makes writing and making notes on the script while it's on the easel difficult without putting holes in the paper, but it's one of the things we gladly trade off to capture the best possible recording. Also, the easel should be tilted such that sound bouncing off of it isn't reflected directly back into the microphone.

FIGURE 6.4

The Studio Ready To Go

Source: Courtesy of Resolution Digital Studios; photo by the author

In trying to find ways to conserve Earth's resources, some have suggested that scripts should no longer exist in paper format. By moving away from the use of paper, we use less energy and protect the environment by cutting fewer trees. After all, in this digital age, there are numerous ways to present the printed word for an actor to read. With that in mind, let's take a look at using laptop and tablet computers for this purpose. I would not recommend using a laptop in the studio for two main reasons: Because of the size and weight of a laptop computer, it's difficult to place the machine at a proper height for the talent to read from. Many music stands won't accept the weight of a laptop and will lower themselves, and there is always the risk of the computer falling to the floor. And then there is the noise issue. Even the quietest of laptops have a cooling fan and perhaps a spinning disk drive inside, and these sounds are easily captured by the microphone and can ruin an otherwise pristine recording. The sounds are very noticeable and annoying in a recording of the voice.

Avoiding the problems of the laptop computer, there is the tablet. Tablets are very quiet, lightweight, and convenient to handle. The price is reasonable compared with a laptop. And there are a number of good-quality, reasonably priced stands available on the market that hold the tablet securely and that are fully adjustable. At present, my favorite is from Pyle Audio. This model clips onto any microphone stand and will adjust in any way that the stand will. Another is from Standzfree. This model doesn't come up high enough to be read from easily. You can find these and other stands at Amazon.com and other retailers.

Many actors are using tablet computers, such as Apple's iPad, the Kindle, or other brands, for script purposes these days. Although they are in widespread use, I would suggest that you think about some of the inherent problems with this technology for a moment. The first hurdle that I see with the tablet being used for voiceover purposes comes from presenting the script on one of these tablets. We have a problem with the type size: Although these devices are convenient, to read a multipage script from one requires the machine being quite close. To be honest, I've found the screen size is a bit too small for our purposes. And then comes the issue of scrolling through the pages. Although a swipe will turn the page, it's not as simple as turning a piece of paper, and, also, the actor can't read ahead to the next sentence, as she could by holding two pages of paper. For me, perhaps the biggest drawback with using electronics for script reading comes with marking the script. Very often, the actor will mark the script where he wants to place emphasis, take a breath, pause for a beat, speed up, and so on. I always put two sharpened pencils on the music stand at the beginning of every session for this purpose, and I know some actors who also carry colored pencils with them to color code certain things on the script. All of this information is very important to the actor, and there are frequently times when the wording of the script changes during the session. "Replace this phrase with this" or "Delete that word," and so on. All of this marking is vastly easier to do with pencil and paper than with software. And, speaking of software, we should mention that most scripts are written with Microsoft Word. The Apple iPad uses Apple's Pages software, and the two are not interchangeable, and so, before the session, the script must be opened and then resaved in the appropriate software. Because of all these reasons, I would recommend against using an iPad or equivalent for script use for the time being, although many actors make daily use of them.

Whether you opt for paper or electronics, we should be cognizant that the copy stand is the actor's workstation, and it should be sturdy and adjustable to be a comfortable fit for anyone who walks into our studio. Because space in a voiceover booth is often limited, the script stand should fit efficiently in the room and not take up undue space, but, at the same time, be flexible enough to be usable over a long period of time and in a wide variety of situations.

SEATED OR STANDING?

Most voice actors (as well as singers and musicians) prefer to perform while standing. This allows unimpeded flow of air from the diaphragm up through the mouth, and the artist can get a more open and resonant sound, as well as being able to move more naturally during the performance. It is recommended that you plan for the talents to be standing when they come into your studio, as the majority of these people prefer that. However, there will be times when a sitting posture is required, for a number of reasons. For instance, if you are recording a long script, standing for the entire time might unreasonably tire the actor, and sitting becomes a requirement. If it is determined that sitting is what is needed, some thought should be put into the type of chair. My first suggestion would be to provide a tall stool, adjustable for a number of differing heights and *noise-free*. This last point goes without saying: If the seat makes creaking or rubbing sounds, the microphone will pick that up and ruin your recording. For this reason, I don't recommend bar stools, especially wooden ones. They are simply not sturdy enough to be used without creating noise. One of the advantages of using the adjustable stool is that the talent can be more or less "half standing." She is not in a completely seated position that cuts off the flow of air, but is not completely standing, either. Thus, she gets the benefits of both positions, without any of the compromises of being fully seated. Also, if this type of seating is provided in the studio, the actor can stand during takes and then take a half step back and relax during the direction from the control room. Because the seat is only a half step behind them, they don't lose their position at the microphone, and all takes will be consistent for editing purposes.

If it is determined that being seated at a table is required, there are some things that you should consider. Remember that the tabletop is highly reflective, and you will be getting some sound bounce from that source back into the microphone. Just as with the script easel, some sound absorption should be provided on the tabletop. Carpet, foam, or some other such material should be used to cover the entire top of the table. The BBC uses an interesting idea for people seated at a table: The tabletop itself is constructed from a thin, perforated material that allows a large amount of the sound to pass directly through the top itself, thus avoiding the reflection problem. If you decide to try this method using perf-board (available in most hardware stores and home centers), be sure to put a big sign on the table that says "Do Not Sit." A piece of half-inch perf-board will not, under any circumstances, hold the weight of even a small individual! And, just as if the actor were standing, we need somewhere to place and hold the script. A script easel should be provided in this circumstance as well, and everything we said about the easel when the talent is standing holds true if the talent is seated: It should be tilted to avoid reflections, padded to be more absorptive, and so on.

MONITORING

It's essential for the voice actor to be able to communicate clearly with those in the control room during the session, and so we need two-way communication. First, the talent needs to hear direction, and for that we provide a talkback, or cue, system. Second, they need to hear playback of a previous take so that they can adjust their performance accordingly. There are two methods available for these purposes. When I started in the industry, the preferred method was to place a loudspeaker in the studio, and all communication and playback were heard in this manner. Times change, and it seems that most voiceover actors are comfortable working with headphones, and through them they hear the talkback from the control room and any playback. You should note that some actors prefer not to wear headphones, because they want to hear their own voice as it sounds in the room (much as musicians like to hear the natural sound of the instrument). These folks will keep the headphones close by and slip them on to hear directions from the producer or engineer and then remove them to perform the next take. The downside of this method is that you can get "bleed" from the headphones into the mic, and the small difference in timing of the audio signals from the actor's mouth and from the headphones can possibly cause phase errors being introduced into the audio signal. Also, if the talent is reading against a prerecorded music track, they must wear headphones.

The upside to having all communication and playback coming through a loudspeaker is that the talent can judge the performance in an acoustical environment and monitor the read in much the same manner as those in the control room. Of course, the engineer must dim or mute the mic signal during playbacks, or feedback will result, and, if there is one thing that will alienate the talent faster than anything else, it's blasting feedback into his ears. Because most actors are accustomed to working with headphones today, it seems that having monitor speakers in the studio can probably be done away with, except in cases such as recording radio dramas and the like, as well as in primarily music studios.

When headphones are used in the studio, an important tool to consider is a headphone mixer for the talent. Yes, we can provide a headphone mix from the console or controller and adjust the overall volume that is going to the talent, but people would like to hear different things while they're performing, and providing a mixer box is a wonderful solution. Imagine that you are recording two actors at once, performing a dialogue piece, with a musical jingle that has vocals coming in at the end. Each of the performers might want to hear the music at a different level in their headphones as they read, and thus individual mixer stations provide a very efficient solution. The stations are available in configurations from mono (volume control only) to eight or more channels (useful for recording musicians, who each might want their own mix of instruments as they perform). The choice of how many channels you decide on is up to you, depending on the type of work done in the studio, but it is recommended that you investigate this monitoring option and give it some serious consideration.

As for the headphones themselves, there are too many options of manufacturer and price point to recommend any one of the myriad models available on the market to you, but I can provide these thoughts: The headphones that you choose for your studio should be comfortable to wear for prolonged periods. Each of us has a different idea of what makes for a "comfortable" set of headphones, and it is very much a matter of personal preference, but please keep comfort in mind when considering what type of headphones to purchase. Also, they should be of the "closed-ear" design

FIGURE 6.5

PreSonus Headphone Splitter

Source: Courtesy of Resolution Digital Studios; photo by Bryen Hensley

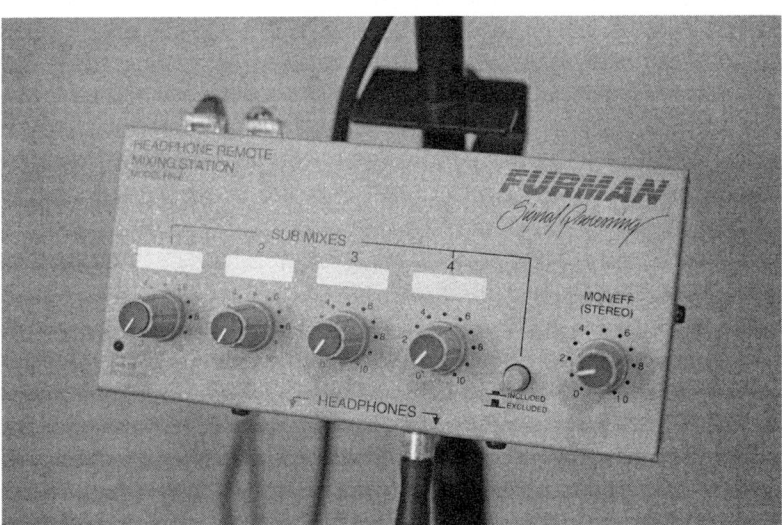

FIGURE 6.6

Furman Headphone Splitter

Source: Courtesy of Resolution Digital Studios; photo by the author

and not "open-air." The open-air type, although lighter and thus probably more comfortable, allows for a large amount of sound to leak out of the headphones and into the room (and, therefore, the mics), as well as allowing outside sounds to distract the talent. A well-sealed pair is much, much better. The headphones should provide a sense of solidity and heft when worn, without being too cumbersome or heavy. Don't use DJ-type headphones; they have an artificial bass-boost built into them, and the talent won't be able accurately to judge what their voice is sounding like. And, when considering headphones for purchase, always keep in mind the durability and ruggedness of the design. These headphones are going to be in constant use, day in and day out, and so durability is a major purchasing criterion. Also take a good look at the ease of repair for the headphones that you choose; repairing headphones is difficult at best, because the wires are so tiny, and some sets of headphones are virtually impossible to repair because of their design. The easier it is to make repairs when necessary, the longer the set will last, and this obviously cuts down on studio expenses.

When thinking about what gear to purchase for your studio, one thing that has always worked for me is to simply ask your user base, that is, the actors, what they prefer. It has always confounded me that manufacturers don't ask their users what would be of use to the final customer, whether that be the design of a new piece of software or where to put certain controls in an automobile. For some reason, the manufacturer thinks that it has all the answers and will dictate to its users how they will use the product. It makes more sense to me to ask the actual users what they prefer and then, if possible, provide that. Of course, your actors (users) may come up with something much too expensive or somehow inappropriate, but, as they are the ones that will be living with your decision, let's get them involved in the process. As I have stressed throughout this section, anything that we, as engineers, can do to help provide for the actor's comfort and performance, the better the result of the session; adequate and comfortable monitoring is certainly one of the things that we can control.

VIDEO MONITORING

There will be times when it will be essential for the actor to see video during the session; as mentioned above, ADR and Foley sessions demand video playback in order for the artists to perform directly to picture. In addition, many television commercials will need video monitoring in order for the VO talent to time their delivery of lines properly. There are currently two methods of displaying video in the studio: One method is to split the video feed, so that it is displayed on both control-room and video monitors at once. The other method displays the DAW's session, which includes the video file. This way of displaying the picture requires either a software or hardware installation, but has a couple of advantages: Some audio workstations have an option to display a larger than normal time counter, which some VO talents find very useful in spotting when to begin delivering their lines and also to give them an idea of whether their read is running too long or too short. Also, the actor can see what the engineer is doing in the control room, and this improves communication if the actor suddenly hears some type of editing taking place and doesn't know what's going on (although you should always explain what is happening in the control room—remember, being alone in the studio can be an isolating experience). Whichever method you choose, be sure to turn the volume of the video monitor off, so that the microphone doesn't pick any program audio that is played into the studio.

The monitor should be placed slightly below eye level for an average-height person and be situated in such a way that, when facing the monitor, the microphone is in one of the best places in the studio for recording audio (remember the "sweet spot" from Chapter 3?). Be aware that the monitor is going to reflect sound, so place the mic appropriately to avoid early reflections coloring your sound. I suggest placing the monitor slightly below eye level so that the talent can move her eyes from the script to the screen with the smallest amount of motion, helping her to keep her place in the script. The ideal placement would enable the talent to see the

FIGURE 6.7

Video Monitoring

Source: Courtesy of Resolution Digital Studios; photo by the author

monitor just over the top of the music or script stand, but this is not possible for all people, because everyone is a different height. A workable compromise for the height of the monitor will have to be found, unless, of course, the monitor is placed on an adjustable stand of some sort. I have made quite useful stands for small monitors by taking the boom off of a mic stand and, in its place, putting a homemade shelf that is bolted to the tilt control that holds the boom to the stand. This solution allows you to tilt the monitor front to back, to avoid glare and to suit individual tastes, and also to adjust the height of the monitor, just as you would a mic boom on the stand. An hour or so in the workshop, some ¾-inch plywood or particleboard, and a mic stand that you don't need at the moment are all you need. This solution won't hold larger monitors, but, if the screen is close enough to the talent, it's very usable and portable, allowing you to wheel it out of the way when not needed, just as you would with any other mic stand. To repeat, the goal is to have the screen situated in such a way that allows for the smallest amount of eye movement to go from script to screen.

The size of the video monitor will depend on the size of the studio, distance of the actor from the screen, and your budget. For critical work such as ADR and Foley, where we are concerned with frame accuracy, the larger the better is a good rule to follow. For simple timing purposes, a smaller monitor will work fine. And one word on getting the picture to the screen: If you are considering using a projection system for your setup, give some serious thought to isolation for the projector. Those things can introduce some nasty fan noise into the recording unless dealt with properly, and, in trying to isolate the projector, you may be subjecting it to heat problems, which will shorten the life of both the projector lamp and the projector's electronics. I don't see any inherent downside to projection, as long as you take the isolation question into account and deal with it accordingly.

KEEP IT ORGANIZED

Remember my admonition to other engineers that I've supervised—"No one notices a clean studio, but *everyone* notices a cluttered one"? This means, not just the control room, but the studio as well. During the session, there are a lot of cables lying around—mic cables, headphone cords, headphone mixer box cables, AC cables, and more. Keep these out of the way of where people will be walking and standing. Instead of having the mic cable just snaking across the floor to the input on the wall, trace it in a neat manner and keep the extra length coiled and out of the way. The same with AC cables and extension cords. And be extra sure that the audio lines and the AC lines aren't running parallel or coiled on top of each other. AC power can transmit into the audio cables and induce a buzz into the audio signal, and so, if they do have to cross each other, make sure that they only cross at 90° angles.

Don't leave miscellaneous gear strewn around the studio; it's a pain trying to navigate a maze or an obstacle course on the way to the mic. Put unused mic and copy stands neatly against a far wall, hang up the (neatly coiled) mic cables so that they're easy to get to if you need another one, hang up unused headphones (or put them on a shelf), and so on. The appearance of your studio is important and it speaks to your professionalism and the respect you show the actors and others who come into your studio. Remember the old adage: "A place for everything, and everything in its place."

ADDITIONAL CONSIDERATIONS

Besides the tools of the trade that are in the studio (microphones, headphones, cables, and so forth), there are some other items that should be placed in the studio for the session. Sharpened pencils (with erasers) should be made available for the inevitable script changes; pens don't cut it in the studio, because, if the need arises for multiple corrections to the script, the ink can't easily be erased and changed. Although a freshly sharpened pencil can easily punch a hole in the page when writing on a piece of carpet that is padding the script stand, it is still much preferred to a worn and stubby pencil. Is there a better way?

Also provide room-temperature bottled water next to the mic and copy stand for the actor. Although an icy drink of water sounds tempting, the cold temperature can cause mouth noise that you'll just have to edit out later, and most VO talents that I know much prefer room temperature when they're on mic. Have plenty of it on hand. As a nice added touch, you can also provide lemon juice next to the water. Nothing cuts through the problems of mouth noise like a quick sip of lemon water, and it's great for the throat. Lemon juice in the water helps to reduce lip-smacking and other mouth noise. Avoid hot beverages: Coffee and tea dry out the voice, and dairy products contribute to the production of phlegm. Another aid that a number of voice actors that I know swear by is apple slices. Be sure to keep these fresh (they turn brown very quickly), and, if providing apple or fresh lemon slices, be sure that you wash them thoroughly before setting them out for the talent. We don't want anyone getting sick from being in our studio! A nice amenity would be a container of throat-lubricating spray (available in music stores or online). Although this is a nice addition, I don't know of a lot of actors who use it during a session. But it's a nice, thoughtful touch.

And please, keep the floor vacuumed and swept; if someone takes snacks into the booth, the crumbs can be unsightly and can contribute to problems with insects and other little critters. Yuck. Once again, no one will notice if you keep everything vacuumed, but everyone will know if you don't, and it will only reflect on your attention to detail (or lack of) if you leave the floor dirty.

The Session

7

Now, it's time for the session. You've arrived early and have made sure everything is set up and your signal chain is working properly. The coffee has been made, and there's plenty of bottled water (and maybe some fresh lemon slices) available. I like to have a box of breath mints set out in the control room, as well as some little snacks such as protein bars and/or hard candy. Because I know the producer is going to have to sit through my editing, I usually will have the day's newspaper on hand to give the producer something to do (and to keep them out of my hair) while I'm working. Oh yes, and I also have made sure that the studio and the control room have been tidied up.

Once the producer and the VO talent arrive, there's usually a short conversation while they go over the script. Make sure you are a part of this conversation. Quite often, at this point, some small changes in wording of the script take place, or perhaps a typo or grammatical error is caught and corrected. You have to make sure that you are working from an accurate script. Your copy of the script will become the master script for the session and must properly reflect any and all changes in wording, as well as pauses and so on that are required. Once the recording commences,

---

INSIGHT: Bryen Hensley, Sound Designer/Mixer, Resolution Digital Studios, Chicago, IL

There are a few qualities that I think a good VO engineer must have, patience being the first. A VO engineer has to work with a lot of personalities: VO talent, copywriters, producers, directors, account managers, etc. All of these people are looking for different things in a VO session. Rarely have I ever had them be all on the same page. The engineer's job is to help guide the clients and VO talent to a successful end where we have a track that everyone loves. Second, the engineer must have an understanding of grammar and sentence construction and word pronunciation. What is written on paper doesn't always translate to the spoken word. And a good engineer should be able to change a word or move a sentence around on a moment's notice, to help the VO talent with the script. Last, the engineer must be technically sound. You never, ever want the equipment to get in the way of a good session, and so knowing your gear inside and out is critical.

I think where engineers get into trouble is when they try to impose too much of their own opinion on a voiceover session. We work for the client; they know what they want—it's our job to help capture that, and trying to put our own creative spin on it just doesn't fly. I'm not saying we have to be totally without opinion, but we have to pick our moments and try to enhance what the client is looking for.

I believe the relationship between the engineer and talent is really undervalued these days. I think a lot of times the engineer is just seen as a button-pusher, when the truth is just the opposite. An engineer and talent who have worked together a lot form one of the best teams you'll see. I, as an engineer, know exactly what that talent is capable of. A good team will have a shorthand with each other that brings a lot to the session. A good engineer will know how to "translate" what a client is asking for to the talent. The talent will be relaxed with an engineer he/she is comfortable with, knowing that the engineer will be capturing/editing that talent and making them look as good as possible. A good team will know when to push for another take, or perhaps take a break to find another approach.

you will take on the additional role of co-producer, and the actor will rely on you to keep them on track. Having an accurate script is a key component of performing this task. Your partnership with the talent is one of the strengths of the session.

We'll have more to say on the partnership with the voice actor later in the chapter, but it's worth remembering that this relationship is one of the most valuable that you can foster.

DOCUMENTATION AND NOTES

Perhaps nothing sets one engineer apart from another more than the quality and quantity of notes that are taken during the course of the session. A careful marking of the script for preferred takes will make your life unimaginably easier and will lead to a smoother-running session. Without accurate session notes and a good take log, the editing process quickly turns into a frustrating and time-consuming experience. You basically have to relive the entire session just to find the proper takes to edit together. Believe me, I've been there, and it's no fun to operate in this way. Also, the producer will think that you are the best because you always have the called-for take at hand within a split second and have comments on what the producer, writer, or whoever else is at the session thought of it at the time of reading. Even if the client doesn't notice you taking notes as the session progresses, he or she will appreciate your efforts (and I have to say that taking good notes and keeping track of take numbers don't involve much effort, especially once you get into the habit). I claim that I can't run a session without a pencil in my hand, and this isn't much of an exaggeration. There are many studios that employ a second engineer or assistant to help the lead engineer in setting up, getting fresh coffee, and so on, and these assistants are often called upon to keep the accurate logs. For a short read such as a radio or television commercial, the notes can be kept on a computer, laptop, or tablet. Other engineers, myself included, prefer to take the notes themselves. Whichever method you prefer, keep in mind that this is an essential component of the session, and so we must be diligent in keeping accurate records. So what kinds of note are we talking about? Figures 7.1 and 7.2 show what I mean.

Each take during a session should be voice slated (the number of the take called out) and recorded as part of the recording, along with the talent's reads. Through voice slating, the producer can keep track of his or her preferred takes and keep their own notes on the session; the talent knows that everything is ready to go, and you are recording; you can refer to these recorded numbers when editing. It's important to get in the habit of voice slating all takes—simply saying "Rolling" or "Go ahead" is not nearly enough. If you are recording your own voice, voice slate each read and keep notes of the takes on your script or a proper log form. This will help you keep track of things, and, when you send your files and the take sheet or a copy of the marked script to the producer or director, they have a good way to reference the various takes as well. (See Chapter 9, Recording for Commercials, for an explanation on setting up a voice slate system in a digital workstation.)

WRENCHMASTER

"Summer Sale: :30 Radio

H19L5430R

MP3 to Sharon-ASAP
" Mark @ wrenchmaster

SUMMER MUSIC UNDER

NARR (:25):

⑦It's that time again for summer vacation with the family, *and* trips to

Grandma's house, ~~and afternoons at the water park.~~ *days* ⑩ But is your car ready for

summer? Come in now for Wrenchmaster's Summer Sale and get a five-

point brake inspection for free, plus 20% off pads and rotors. And don't

forget a summer oil change – with a new filter for only $22.95.

Make sure that your summer travels are safe and problem free – only at

Wrenchmaster! ~~It's summer~~time – *relax*!   *13 B*

TAG (:05):

⑧Sale prices good June 15th through July 20th. Not good with other

offers or sales. Not valid in Arizona or Louisiana.

FIGURE 7.1

30-Second Radio Commercial

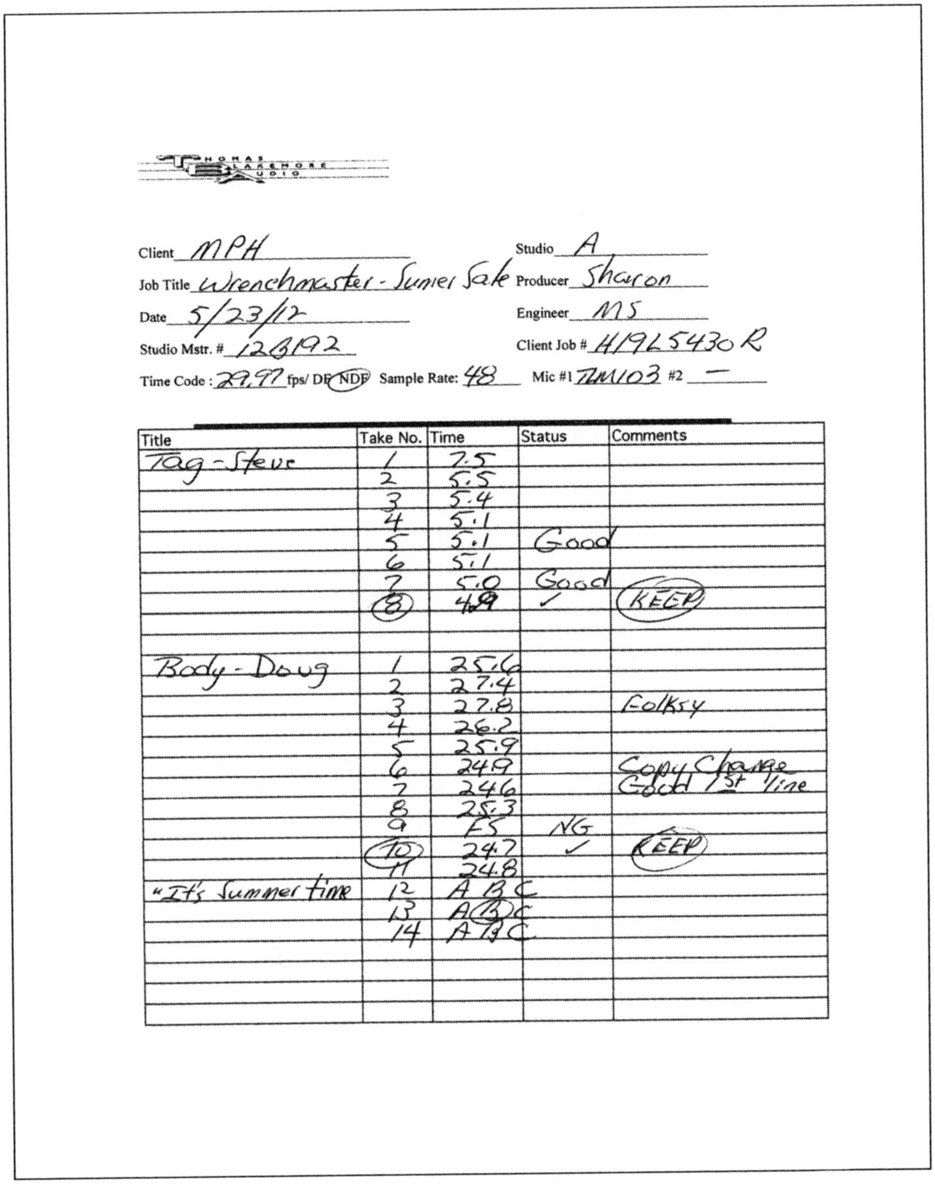

THOMAS BLAKEMORE AUDIO

Client MPH
Studio A
Job Title Wrenchmaster - Sumer Sale
Producer Sharon
Date 5/23/12
Engineer MS
Studio Mstr. # 12B192
Client Job # H19L5430 R
Time Code : 29.97 fps/ DF NDF Sample Rate: 48 Mic #1 TLM103 #2 —

Title	Take No.	Time	Status	Comments
Tag - Steve	1	7.5		
	2	5.5		
	3	5.4		
	4	5.1		
	5	5.1	Good	
	6	5.1		
	7	5.0	Good	
	8	4.9	✓	KEEP
Body - Doug	1	25.6		
	2	27.4		
	3	27.8		Folksy
	4	26.2		
	5	25.9		
	6	24.9		Copy Change
	7	24.6		Good 1st line
	8	25.3		
	9	FS	NG	
	10	24.7	✓	KEEP
	11	24.8		
"It's Summer time	12	A B C		
	13	A B C		
	14	A B C		

FIGURE 7.2

Take Sheet

Figure 7.2 is an example of a log for a radio commercial. Each take is numbered, and its time length is noted. There are also comments on the various takes, such as which are the preferred takes, where the wording was changed to help get the read into the allotted time (30 seconds), and other comments as needed. (Again, refer to Chapter 9, Recording for Commercials, for additional comments on auto-numbering your recordings in a digital workstation to help with quickly navigating to the take(s) that the producer may request.) But it all starts here, with a good take sheet. You can easily create a very usable take sheet and customize it to match your needs using any number of software programs, and, if you customize your own, it will become even more valuable and useful to you, as it will reflect those items that are of the most concern to you and your way of working.

From this take sheet, we can see that a total of fourteen individual reads were performed of the body of the spot, and there were eight takes of the tag. The preferred base read is take number 10. The base read is that take on which the majority of the performance will be based. Then, from take 7, we would like to grab the first sentence, "It's that time again for summer vacation with the family and trips to Grandma's house," and insert that into take 10. Now, on takes 12–14, the actor did what is known as a "series of three" for the ending line of "It's summertime—relax." Each of these series of three reads is slated together on one take, and the notes read, "A, B, C." The preferred read is number 13B, and this will also be edited into the base take of number 10. Without these notes on the take sheet, we would never remember which of the takes are to be edited together for our final product.

Some producers will record read after read, looking for that "golden take." Others, with the skill and input from a knowledgeable engineer, have the ability to stitch together the best parts of multiple takes to form a coherent and powerful whole. We see a bit of that process with this commercial.

There is space to note a number of other things besides take notes on this log sheet. Look carefully at what else is documented on this form. Especially for commercial work, but equally important for any kind of voice recording, we should first of all note what microphone was used for the session. Also, in a multiroom facility, we have to know in which room the session took place. Quite often with commercial production there are redos and inserts done after the fact, and, if we expect a proper match of recording tonality, these parameters must be known and matched. The easy automating of session parameters such as EQ and compression in a digital workstation simplifies this matching considerably, but still these notes are vital to ensure consistency. The number of "fix" sessions that accompany commercial production often bewilders the novice or outsider, but, with this type of production, there is often a considerable amount of money at stake, and the producers go to great lengths to ensure that everything is exactly right. And, of course, we should note the producer's name, the date of the session, and the session code number for reference purposes later on. I will have more to say on the importance of proper naming and session code numbers later on in this chapter.

This type of form works well for those projects that are timed and relatively short, but, for a longer form of read, it doesn't meet our needs. If you have a script

InterPark   Page 1 of 6
Customer Service   5/11/05
"Best Practices"   Draft G

<u>SCENE</u>   <u>ANNCR:</u>

1.   The demand for parking in urban centers is outstripping the supply of
      spaces. And that's the forecast for the foreseeable future.

2.   Because even as new parking facilities are developed, America has
      changed the <u>kind</u> of demand it has for parking.

3.   Urban growth creates more high-rises and a higher density of
      population. This congestion creates more customer demand for
      parking.

4.   Today's cities are multi-use environments that combine work,
      shopping, tourism, entertainment and residences.

5.   And city parkers today having a growing need for parking 24 hours a
      day, 365 days a year.

6.   Today, InterPark leads the industry in parking innovations and
      customer service. This video will share our best practices to make
      your job as smooth and successful as can be.

<u>GORDON HALMEYER INTERVIEW</u>

7.   From experience, we know that success depends on executing every
      day. This execution is grounded in our pledge to provide a clean, safe,
      courteous and efficient parking experience to every parker.

<u>ANNCR</u>

8.   So, where do we start? First, remember that your customers won't
      know that we operate more than 300 parking facilities nationwide . . .
      that we manage over 135,000 parking spaces across the country or
      that our garages process more than 2.5 million transactions a month.
      With that many transactions, you have to be ready for anything.

FIGURE 7.3

Long-Form Script

Source: Courtesy of InterPark Corp.

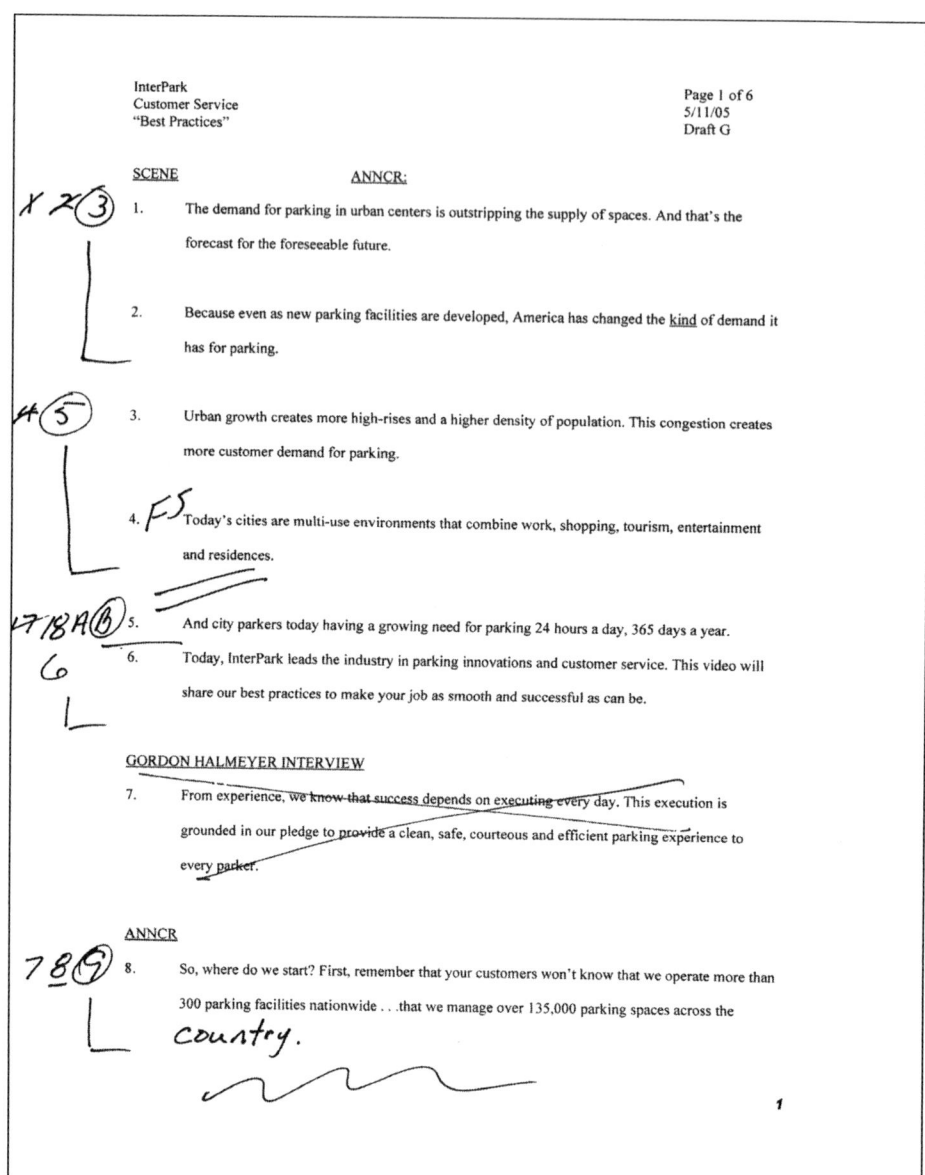

FIGURE 7.4

Script with Edit Markings

Source: Courtesy of InterPark Corp.

that runs to multiple pages, trying to use a take sheet to keep accurate records of where each take begins and ends becomes almost useless. Take a look at a single page from a long-form script—in this case, a training video for new employees on the business practices of a corporation—in Figure 7.3.

With this type of project, it is not practical to try to keep track of the beginning and end of the good takes with a take sheet. In this case, we make notes directly on the script. With a commercial script, such as in Figure 7.1, although we could simply make notes on the script, it is much more efficient to use the take sheet, as we can keep accurate comments on each take as we go along. If we try to keep track of the reads directly on the script, we don't have adequate space to make comments and note word changes, and so on. However, with a long-form script, marking the script directly makes more sense and is much easier to make sense of. Figure 7.4 shows the same script marked for proper takes after the session.

Here, you can easily see which are the preferred takes and which might also be usable. This is one of my scripts, and I have developed my own shorthand for marking scripts that is very logical to me, but can also be picked up easily by anyone that I might pass this recording on to for editing. Here are some explanations:

- A take number that is crossed out is an unusable take, for whatever reason.
- If it's a good take, but not the preferred take, there is no marking on the number.
- If I find something I like in the take, I underline the number.
- The preferred take number is circled, and a line is drawn to indicate how far it runs.
- "FS" before a line indicates a false start, but not a new slate number.
- The squiggle line at the bottom of the page indicates there is some sort of noise that must be edited out (perhaps the sound of paper shuffling or a throat clear).
- The double lines between paragraphs indicate a long pause that must be tightened up.

Also note that, after the actor had read all the way through the script, we went back and rerecorded an insert line for a change in emphasis or pronunciation that we picked up after the initial read through. Furthermore, because the end of the final line on this page continued onto the next page, for ease of reading the line is written in full on the bottom of the first page. All of this is clearly marked and gives the editor (whether yourself or someone else) a road map to follow to put everything together. You can begin to understand the problematic nature of attempting to efficiently edit this all together without adequate notes having been taken.

So, we see that these simple shorthand markings all mean something to the editor and help in the process of putting together a complete read. I've developed these markings over the course of my recording and editing career, and there is certainly nothing magical about them. Feel free to come up with your own shorthand, but it should be consistent with everything you record and easily understood by anyone who picks up the script and is going to perform the editing. Anyone who has worked with me for more than about an hour knows immediately what I am attempting to convey with these markings and can quickly and efficiently complete the edit.

THE CO-PRODUCER ROLE

In running the session, you will find yourself in the role of co-producer. This can be a delicate political dance for you, the engineer, because oftentimes the producer thinks that he or she has all the answers and doesn't appreciate your input. Remember that this is not your project; you are here to help all parties arrive at the outcome that they want. Don't ever attempt to take over the session. With experience, you will develop a sense of when to speak up and how to do it in an inoffensive manner. It's true that some producers only want a button-pusher; if this is the case, accept it and push the buttons. However, those producers who are truly after the best possible product will come to appreciate your concern for their work and, with time, come to depend on you for your insights and suggestions. If you are working with a producer with whom you are not familiar (and who hasn't worked with you very often), you can couch your suggestions in non-threatening language, such as, "Just thinking out loud here, but what would it sound like if we said . . .?" This is much easier for a fragile ego to take, rather than a firm declaration that you have a *much* better idea than the producer or writer. Again, this is part of developing the sense of when and how to speak up, in a collaborative spirit, rather than seeming like you are trying to take over their session.

OK, remember my comments on multitasking earlier in this book? This is where we find out how good at multitasking we really are. We are, of course, paying attention to all of the technical issues that are flowing past us, but we are also *listening* very closely to how the performance sounds. I once heard a well-respected engineer describe sitting in a session this way (and I'm paraphrasing here):

When doing a session, you have to pay attention to the minutiae of what is happening; you have thousands of pieces of information flowing past you every second, in real time, and you have to be aware of each of them. But you can't lose sight of the macro. What is the big picture? How do all of these little bits fit into that?

I find this distinction between the micro and the macro very interesting and appropriate for us to consider. Of course we are paying attention to all of the technical details of recording, and, if you hear something technical that comes up, point it out and make sure that it is corrected. After all, the technical issues are your main concern. This includes anything in the talent's read, such as popped Ps or extraneous noise from jewelry or paper shuffling. Again, correct these immediately as need be (suggest that the actor removes her jangly bracelet, make a small mic adjustment to avoid the popping, etc.)—don't let these types of thing linger take after take.

Pay attention to grammar and sentence structure—if it hits your ear as odd, speak up. Watch for changes in tense, gender changes mid-paragraph, the proper use of grammar, and any unclear writing you detect. If it sounds unclear to you, it will probably be unclear to the listener, as well. After all, you have a script in front of you, and the audience doesn't. Oftentimes, writers write for the page (the eye) and not for the ear; be very sure that, in following along on the script, you keep your ears open and make sure that everything is making sense.

Something else to pay attention to as the reading continues is how everything is going to be able to be edited together once the actor finishes. Be aware of tempo and voice timbre especially. As part of looking ahead to the editing process, don't change your levels or EQ settings mid session. This is what the process of setting up and getting initial levels and settings is for. Of course, there will always be times when you have to adjust the recording volume during the session, but, if you do, try and make as slight a change as you can get away with, or at least be aware that, in editing and mixing, you'll have to make adjustments.

All of this is part of the co-producer role, and there is much more to it that you should be aware of. For instance, the producer may request a certain piece of music or a sound effect; if you have a better idea, don't hesitate to bring it up, but do so in a respectful manner and offer it as an alternative suggestion—something that your producer hadn't thought about. Your idea may get shot down, but don't take it personally. After all, everyone is working toward the same end, and that is a great product.

WORKING WITH THE VOICE TALENT

When making comments and suggestions to the talent, be as specific and as brief as you possibly can. The worst direction any actor can receive is, "No, that's not right. Try it again." What is the actor suppose to do? Are you asking them to try endless variations until they stumble across something that works? Ask yourself, "Why isn't this read working?" What suggestions can be made to move the read closer to what you should be hearing? *Be specific.* You don't have to give a seminar on the subject; go directly to the point. The same goes when discussing the read with the producer: Be specific and be brief. Directing is a special skill, and it can take a long time to develop, but keep at it; you'll find that you get better at it, and the actors might give you some pointers about it if you ask. After all, they appreciate good direction more than anyone, and anything they can do to help the process they'll be glad to do. Now, there will be times when you, the engineer, and the producer will be discussing things in the control room that aren't directly being addressed to the talent. Perhaps you are discussing whether the grammar in a sentence is correct, or if a word should be changed to help the clarity of the message. Here's a tip that I've found actors really like: When these types of discussion occur, I casually rest my hand on the talkback button and let the talent hear what's being discussed. Sitting in a glassed-in room all by oneself can be a lonely experience, and actors are by nature an insecure lot. If you leave them out of the loop on these discussions, they often jump to the conclusion that you and the producer are talking about them and what a terrible job they're doing with this read. Most actors are astute enough not to let on that they are eavesdropping on the conversation, but, if they are aware of what is being discussed, they will feel much better about the

INSIGHT: Harlan Hogan, Voice Actor, Chicago, IL

When I'm booked at a professional recording studio, I have rarely been asked if there was a particular microphone I'd prefer, and I'd never suggest one unless asked. That said, I like the popular Sennheiser 416 for—and *only* for—in-your-face broadcast or promo announcements (i.e., political attack ads, movie trailers, TV bumpers, etc.). The 416 requires very disciplined mic technique and it's usually not flattering to female voices. For long form, it's hard to beat a U87 or one of a number of less expensive mics modeled after the 87's sound.

Aside from recording chops (a given), a successful engineer will possess great people skills and the ability to guide and manage a session, all the while working with clients who possess everything from very little to vast experience and with voice talents with similar ranges of experience.

To help us give our best performance as actors, allow a few minutes for us to become acquainted with the producers and their expectations for the project before shuttling us off to the recording booth. We all appreciate a clean, brightly lit room, individual headphone gain control, perhaps a glass of water or a cup of coffee, a highlighter and a couple of pencils on the copy stand, and a box of tissues nearby. That's about it, and we're ready to go.

The best engineers know how to work *with* the writers, producers, and clients and contribute positive ideas to the session, without *becoming* the producer. Over the years, I've seen that happen several times. Once, the engineer started arguing with the writer over the reads. Another time, as I was leaving a session, I overheard a relatively new engineer tell a client he could fix my reading in post. Later, I called him and—though seething—explained that the time to fix a talent's reading is during the session, and that his remark cast doubt on my performance and my professionalism. Truth is it was equally insulting to the client, who had been very pleased with the recordings until the engineer crossed the line and became a "producer."

I *love* it when all the parties involved—including talent—collaborate during a session, but that is totally up to the clients. If asked, I'll contribute my two-cents' worth, but, otherwise, I stick to reading the script as directed. I also appreciate the engineers who are aware of how easy it is for a performer to get a bit paranoid, watching through the glass as the creative team is engaged in an obviously heated conversation. "Is that about me?" is a natural reaction. So the engineer simply opens up the control-room mic. "Whew, they're discussing how awful the script is, not how awful I am." Talent of course, should *never* reveal that little collaborative secret.

As for preparation and helping the voice actor feel at home, there's nothing nicer than a supportive engineer who's got the mic at "Harlan Height" when I walk in the door. The great ones subtly work with the talent when they need some help on direction or pronunciation. I'll never forget the day I went into your booth, looked at the script, and saw a name of a candidate running for state senator that I'd never heard of, much less knew how to pronounce. I didn't want our producer to know, so I drew my index finger across my throat—the universal sign between talent and engineer to "kill" the microphone. You cheerily came in to the booth and pretended to adjust the mic's height, even though it was already at Harlan Height. "What's up Harlan?" "Tom, how the hell do I pronounce this guy's name?" "Oh, it's Bah Rock—Bah Rock Oh Bahma." "Thanks!" "Yeah, he was in here again last week; pretty cool guy, word is he might run for President someday." "With that name? That'll be the day . . ."

situation. And, by having my finger on the talkback button the whole time, if something comes up in the conversation that I don't want the talent to hear, I can unobtrusively let up on the button, and the producer has no idea what I'm doing.

Nice pointers, Harlan. And Harlan brings up another interesting subject—that of nonverbal communication. Back in the old days, communication systems weren't as sophisticated as now, and, in live radio, to have someone suddenly talking into your ear while you were reading could really ruin the concentration. So, a system of hand signals came into being, and these hand signals are just as useful today. A number of voice actors I've worked with have a series of signals to communicate with the engineer, sometimes based on the old radio method and sometimes completely new and original. Pulling your hands apart is a universal sign for "Stretch," i.e., we have more time to fill. Rotating your hands in a circular motion around each other means, "Wrap it up," or we're running out of time. The finger across the throat means, "Cut," and there are plenty of others that I've seen. One actor has a signal where he taps his wrist with his finger. The meaning is, "What's this guy's name?" In this case, I'll direct a comment to the producer and make sure that I include his name in the comment: "Uh, Bill—what did you think of that last read?" Of course, my finger is on the talkback button when I say this, and the talent is reminded of the producer's name, which he had forgotten. By the way, if someone comes into the studio that I'm unfamiliar with, I'll write their name on my copy of the script, because I am not particularly good at remembering names. This comes in especially handy if there is a group of three or four people who come in. I'll write their names in the order that they are sitting in the room, and this way I can address each by name (and remind the talent as well). There are many more hand signals, some subtle, some not, that can be used during the course of a session, and working in this way really speeds up the communication process and keeps everyone involved in the loop. By the way, I once worked on a film set where a number of the crew members knew, and utilized, American sign language. Great for when the camera's rolling, and conversation between the crew members can't be tolerated, and especially effective across long distances, where a whisper wouldn't be heard. It's a really good idea, and one that I would recommend to anyone working in this particular field to consider.

Communicating on the talkback, hand signals, whatever you have to do to keep a clear communication path between all parties, all of these and more are keys to the successful session. One other thing to remember is to keep the talent in the loop at all times, and let them know what you are doing at each step. Nothing is more disconcerting than waiting to hear the next bit of direction and suddenly hearing some editing going on or one of your reads playing over and over in the headphones. If you, as the engineer, want to check one of the takes for some reason, simply open up the talkback and announce, "Playback of your last take, Sam." Or, "Hang on— I want to try an edit; relax for a minute." You can even invite the actor to take a short break and come out of the booth and stretch a bit. Above all else, please let them know what you're doing, just as you did when you opened up the talkback mic so that they could hear the conversation between you and the producer. No one likes being kept in the dark, and a simple explanation is the best way to keep the team running smoothly. And something that shouldn't be overlooked: The vocal booth can be a lonely place with only one actor in it. Let the talent know that you are still out there!

You should develop the skill of turning the pages of a script silently, just as the actor does. One of the things you should be listening closely for is the sound of paper being shuffled—this is totally unacceptable in a finished recording. However, if you turn the page and make a lot of noise doing it, you won't notice that the talent also made some noise in turning her page and you will miss correcting it, either by cutting it out or having the actor read the line again. Once they leave the studio, it's too late to correct.

If you have information to pass to the talent that you don't want the producer to hear, announce to the producer that you have to make an adjustment to the microphone (they won't know why and generally won't care—this just shows your attention to detail and how good an engineer you are). You can then mute the mic, go into the studio, and pretend to make an adjustment while giving the information. I've had situations where the producer was notoriously difficult to work for (for a number of reasons), and, if the talent hadn't worked for this particular client in the past, they wouldn't be prepared for what was about to transpire. I might see the actor becoming flustered or frustrated, so into the booth I would go, and tell them that this client was extremely loyal to the people he worked with, but often made what seemed like unreasonable demands for doing things over or changing the wording in a script all day long. If the talent has this information, they can relax, because they'll realize that it isn't their fault that the session is going the way it is, and it's just the standard operating procedure of the producer in question. Just be doubly sure that you have turned the microphone off before you have this short conversation!

Now for a touchy situation: What do you do if the producer realizes that the actor who's in the booth and reading is the wrong choice for the project, and they have to fire them? Well, ideally, this should be part of the producer's responsibility.

INSIGHT: Doug James, Voice Actor, Chicago, IL

As for microphone choices, I have a couple of favorites; I've always liked the Sennheiser 416, and the other is the Neumann U87.

The engineer should have great people skills and, obviously, a firm grasp of the digital medium. If the engineer isn't a musician, they should have a musician's sense of timing, flow, and dynamics at the very least. The engineer should be attentive all the time. Know that sarcasm doesn't work with most talent unless you are their close friend, and sometimes that doesn't work, either. The engineer will make sure the room temperature is comfortable, and always have plenty of room-temperature water available. And, if the engineer is providing copies of the script, please make sure that the copy is double-spaced!

In the session, we're all a team trying to get the right pacing and delivery. Any help that we can give each other is great. I always appreciated the engineer allowing the client and the talent to keep an open dialogue; it's much easier to get the direction from the source. And, if the talent wants to do a "series of three," let them. The talent should also be willing to reread a paragraph to help the engineer keep editing to a minimum.

But I've been thrust into the situation of having to go into the studio and inform the talent that they are just not working out, and the producer is going to have to find someone else. Tough place to be in, isn't it? The better your relationship with the talent, the less uncomfortable this becomes, but it's never an easy task. Usually, this happens because the client picked the wrong talent to begin with, and it's not an issue of the actor being incompetent. In this case, the straight, simple truth is the best, and a professional will accept the news, take their session fee, and go on with their life and career. In other words, they don't take it personally. It happens to everyone at some point. Trying to cover up the situation by saying, "OK, great read—we've got what we need; good work. See you around." rarely works, because the actor almost always finds out the real story sooner or later, especially if the project is a commercial job, and they hear someone else doing the spot on the air. Just explain that a mistake in casting was made, and, no matter how good the actor is, they're just not what the producer is looking for. Maybe the actor is female, and, in hearing some reads, the producer comes to realize that a male would work better. Or maybe someone younger, or older, or a lower voice, or . . . on and on it could go, but you get the general idea. Last word on this subject: Tell everyone the truth and you'll avoid hard feelings in the future that could jeopardize relationships. Remember, these relationships are your lifeblood as an engineer.

SOME TIPS AND TRICKS

Sure, I believe in rules . . . sure I do. If there weren't
any rules, how could you break them?
(Leo Durocher, baseball player and
manager; Hall of Fame, 1994)

DAW software comes with a wide variety of processing programs—plug-ins. There are also many more available from third-party suppliers, such as Waves. They can be automated and recalled at any time. The plug-ins make available to us a large number of effects, and, by combining different effects together, we can achieve some truly amazing-sounding audio; from the voice of an alien creature to an explosion created from the sound of a pop-top can, plug-ins are incredibly adaptable to a wide range of applications. But what happens if you still need a sound that is unique and can't be arrived at with plug-ins, no matter how you try to tweak them? It may be time for a little do-it-yourself processing.

There's a huge choice of objects that change the tonal characteristic of a voice in unexpected ways that you might experiment with. Have someone speak into one end of a cardboard or PVC tube and place the mic at the other end. Variations on this are almost endless: Try tubes of different diameters and lengths; with cardboard, try putting a 45° or 90° bend in the tube; place a small microphone (perhaps a lavalier mic) part way into the tube and try moving it closer to or further from the actor's mouth—this can also be done by cutting a narrow slit in the tube and suspending the mic from its cable, down inside the tube. Roll up a piece of cardboard into a megaphone. Record while the performer speaks into a bullhorn from across the studio. Try using two mics to record someone speaking while wearing a full-face motorcycle helmet with the face shield closed—a lavalier inside the helmet and another mic outside, pointing toward the face shield—and then combine the two signals by varying amounts until you get to the sound you're after.

It may look strange, but have the person talk across the mouth of an empty five-gallon water bottle. Or drop a small mic down inside. Try varying the distance from the mouth to the jug's opening.

If you have access to a studio with an electric organ (such as a Hammond B3), send the voice signal through its Leslie speaker cabinet. The spinning speaker can work wonders on the voice—John Lennon and George Martin used this effect on a couple of Beatle's songs.

Electronic Musician magazine and others occasionally run articles on strange recording techniques (aimed at the music recording audience, but they work equally well on the spoken word). The Internet also has a number of sites devoted to off-the-wall ideas.

Think outside the box and experiment. You can never tell what it is that will give you what you're hearing in your head. Some other ideas to consider and play with are:

- Tape a contact mic or PZM to the studio glass.
- Record under water: Use an underwater speaker (a hydrophone) and a waterproofed mic (you can tightly tape a condom around the mic body—just make sure that this is a mic that you don't mind getting wet, if the waterproofing fails)—and not a tube microphone, as these could produce a lethal electric shock!
- Try out truly terrible-sounding mics—visit a vintage toy store or eBay and latch onto a "Mr. Microphone" toy.
- Record the voice coming through a blown speaker.
- Use guitar effects pedals ("stomp boxes").

These are only a small selection of things that you might want to try out. Your imagination will take you to places that I can't even imagine, so have fun!

Often, we might want an effect on the voice that can't be duplicated in the mixing stage—the sound of someone walking into a room while speaking or sounding like they're wrestling with someone, or things of that nature. In this case, we can direct the actor to simulate these situations by a variety of means. One thing to keep in mind, however, is to keep these effects to a minimum, or you risk distracting the audience by having them pay more attention to the effect than the content that they are supposed to be listening to. Most of these techniques were developed in old-time radio and have been around seemingly forever, but it's amazing how few of them are used these days. For some reason, it's as if we have forgotten the easy ways to do things, as the technology has become more sophisticated. But here are some things that you might want to consider and experiment with.

Let's say that a character is supposed to be walking into a room as they deliver a line. You could pull down the level in the mix and then increase it as they get closer, but this approach fails to give the full illusion of them walking in. There are two ways of simulating this effect. One is to have the actor simply turn their head 90° as they begin to deliver the line, and then gradually (over the course of the line) turn back to face the mic full on. Not only will there be a change in volume as they turn into the mic, but the EQ will naturally shift as well, which is what happens when someone is at a distance from you and then gets closer. There is a drop-off in high frequencies at a distance, and this will be replicated as the actor moves from an off-axis position at the microphone to a more on-axis one. Another way of creating this effect is to have the actor back away from the microphone at the beginning of the line and then gradually move closer to the mic. No more than three or four steps will give a nice "walk-on" effect. With a bit of reverb added for the distance portion of the line, and then decreasing reverb as the actor gets closer to the microphone, this technique creates a very convincing illusion of space and movement. As with any audio effect, a touch of reverb here goes a long way: No need to overdo it. One thing to be careful of when having the actor do this is the actor getting too far off mic. The sound of being off mic is very distracting to the listener and most times is interpreted as a mistake. Audio perspective (distance effects) is a fine line, and your judgment will determine how far to push this technique. Most times, a small amount of change from off axis to on axis sells the idea of perspective, and you won't have to go to extreme lengths. Also, don't use this for an extended amount of time, as, again, this can be heard as a mistake.

If a character is supposed to sound out of breath when speaking a line, try having the talent run in place for a short period of time before you drop into record. There is no way to fake being out of breath and have it sound convincing, and this technique is very effective. Just make sure that the actor doesn't go overboard and unduly tire him- or herself. If there is supposed to be a struggle going on in the script, have the actor move about a bit as they read the lines. The quick on-and-off-axis sound that you get is quite natural sounding, and the actor's breathing also becomes faster, which is also natural sounding in a real struggle. These methods of achieving a "real-world" sound are very effective if done at the right time and in moderation and have been time tested over the years. There's no rule that says that a voice actor has to stand perfectly still in front of the microphone at all times, and, if the script calls for movement of some sort, experiment with having the talent actually move. Along these lines, I've occasionally resorted to location audio recording techniques in the studio. If called for, I'll set up a shotgun microphone on a boom pole and follow the actor around the studio, as they move about and act things out physically. Just be careful of overhead lights when using a boom pole indoors! If this is done, the actor has the freedom to move as much as they need to give the right performance; the downside is that editing a number of takes together can become problematic because of differing mic perspectives on the talent, as well as the sound of footsteps, if they become audible.

If one side of a conversation takes place on a telephone, you could filter the recording while mixing. However, I once worked at a studio where the chief technical engineer took a real telephone, removed the earpiece, and then wired the output of the phone mouthpiece to an XLR cable that could be plugged into the recording console. Presto—a convincing-sounding telephone that also gives the actor the feel of really talking on a telephone. The downside of this method is that the sound that you get is the sound that you get—it doesn't vary if you need a different-sounding telephone, but, if that's the case, you can always revert to filtering the voice after the fact, having recorded the lines flat. By recording the actor with a conventionally placed microphone at the same time, it is possible to adjust the amount of telephone sound versus "natural" sound, and this has worked out very well in a number of situations.

What do you do if a script has to be read in a limited amount of time—for instance, when recording a 30-second commercial—but the talent is having a problem fitting all of the words into the allotted time? In most instances, the talent is going to feel rushed and begin pushing harder and harder to get the words in, often raising their voice in the process. Have the actor relax and give a softer reading. Surprisingly, many times, this will result in being able to fit more words into the time window. The more we strain and the louder we speak, the slower our speech becomes. So, before starting to cut words, try pulling back on the delivery.

If the script calls for a dialogue between two or more characters, try and have all of the actors in the studio at the same time. They can play off of each other, instead of having to imagine the other half of the conversation, and the resulting performance will be much better. Also, if both actors are reading at the same time, it drastically cuts down on the amount of editing that you'll have to do to put everything in the proper order, but the most important reason by far is the performance that will be given. This is especially true for comedy scripts, but dramatic content also benefits from the talent actually acting together as a team. After all, they're actors, and this is what actors do.

If the script contains proper names or unfamiliar words, request that things are spelled out phonetically. Now, not all names need be spelled out; for instance, Smith or Brown or Jones is already about as phonetic as it is ever going to be. But place names or the names of unfamiliar products and procedures can be confusing and can cause headaches for even the most accomplished actor. This is especially true for medical or pharmaceutical scripts, where we're dealing with the names of drugs, parts of the body, or specialized procedures that those who practice certain professions are familiar with, but the general public is not. Even when spelled out phonetically, the talent will probably slow down significantly to figure out how to pronounce the word for the first couple of reads, until they become more and more familiar with saying it, but this is better than repeating the pronunciation over and over for the actor using the talkback. This is one of many good reasons to have the script printed double-spaced, by the way. It gives the talent room to make notes on the pronunciation directly above the word in question and saves them having to search the margins of the page for a note when they need it.

When recording a long-form script, if the actor makes a mistake and has to reread a line or start a paragraph over again, suggest that they begin a sentence or two back in the script and take a "running start" at the edit point. This will result in a much smoother edit, because the read will be in rhythm and won't jump nearly as much in volume compared with the line immediately preceding it. This "read-in" technique is also useful for dialogue scripts: Have the first actor deliver their previous line, so that the second actor has something to react to, instead of delivering the line cold. It just sounds much more natural. Be sure to mark your script accordingly, to avoid including repeated words in the completed edit.

There are many, many more of these types of tricks and techniques in use, and every engineer has their own favorites. Learn as much as you can by observing, listening, and speaking to other engineers. The more ways we know about to get that great performance, the better and more creative our work will become.

DOCUMENTATION (PART II)

Once the session has finished and the talent and producer have left the studio, your tasks are not finished. There is further documentation that must be accomplished. This is a vital element that keeps the studio business functioning and greatly reduces the chances of mistakes being made that could adversely affect your professional future. The first of these tasks should be documenting the time it took to do the session, which is the basis on which most billing to the client will be done. Recording work can be billed either by the hour or on a "bid" arrangement, whereby I agree to do the work required for "X" amount of dollars, no matter how long the work might take. Either way, an accurate time record is essential for the records. Every facility and each independent contractor has their own preferred method for this documentation, but, whatever system you might devise and use, make sure that what you are reporting is accurate, honest, and fair. The easiest way in the world to lose a client is to try and charge them more for the job than was actually done. They know how long they were at the studio and how long your editing and mixing of their project should take, and they won't hesitate to call you on it if you overbill something. So, make sure that the records you keep and report are accurate. Also, if you did this project "on bid," you can look back at past projects to project how much you should be charging on the next one.

Then comes the process of archiving and logging all materials that were used in the session. Regardless if you are recording on analog tape, digital files in a workstation, or engraving the words in stone, *all* materials must be archived (also known as backing up your files and materials) and properly logged, so that, at any time in the future, you can see exactly what you did and can, in a reasonably short period of time, put your hands on everything having to do with the session. The goal is to be able to recreate this session years later if need be, and accurate archiving of *all* materials and notes can save you if any controversy (not to mention legal action) ever comes up in the future. I can't stress this aspect of the session process strongly enough.

The key to the entire process of archiving and documenting is the session number, also known as the vault number, tracking number, and many other names. And the secret of the numbering system is to make it consistent and understandable. There are probably as many different systems as there are studios and independents, but here are a couple of simple and easy ideas. One good way of keeping track of things is just a consecutive numbering method: For instance, start at 0001 for the first session you do and increment the numbers up for every session. Although this method certainly keeps track of things, it doesn't tell us much about the individual session. We could use a dating and numbering system; if the session is done in June of 2014, we could start our session number with the year, month, and then the individual number. So, in this case, our session number might be 1406001. Here, we can see the year and month the session was done and have a unique number for each of the sessions done that month. You begin to get the idea. The session materials are then put into the library and filed under this number, all billing refers to this number, and all client correspondence asks that the client refer to this number when inquiring about a past project. We'll get to what goes into the library in a minute. If you set up a searchable database to keep track of your jobs, you can search for information on the session by client name, date, session name, and session number. I have had instances where a client calls and says, "I did a session for this particular product about two years ago—can't remember the exact date—and we used a piece of music on that spot. I want to use is again; can you dig it up and send it over to me?" Well, where to start? With a good archiving system, this search should be relatively painless and the piece of music found quickly. Once again, it shows that you are on top of your game, and the client works with you for this reason.

All session logs and notes should be saved into the library, along with your original tapes or digital backups and archives. In the old days, we were dealing with 10-inch reels of 2-inch tape, sometimes multiple reels for a big project, and those things were a bear to store. They took up so much room and were very heavy. Nowadays, we can slip a DVD of a session backup into the archives much more easily and be done with it (see Chapter 14, That's a Wrap, for a further discussion of this part of the process). Storage becomes a lot easier! I've seen many ideas on how to store all of the materials—backups, scripts, logs, notes, delivery information, and so on—and there are many possible solutions that might work for you. Some facilities use standard-sized cardboard boxes, others use accordion manila folders, but, into the container goes *everything* that was generated during the session. This includes the script, take sheets, music licensing forms, lists of what sound effects were used, delivery notes, and receipts—in other words, everything that documents what happened in the studio during that session. If the client comes back two years later and wants to make a revision (not as uncommon as you might think), you should be able to put your hands on everything that you will need. The session code number goes on the container, along with the client and product and the date of the session; if there are so many materials that they don't fit into one container (not an unusual situation for a long or complicated project), the box, folder, or whatever you're using for storage should be clearly marked "1 of 3," "2 of 3," and so on. For instance, the script for a video game can run to hundreds of pages and will obviously take a lot of room to store.

The message here is: Keep everything! In my files, not only do I have the more obvious materials, but also little notes on how many copies of the mix and in what format were sent to what people and by what method of delivery, and anything else that I can pull out and use for documentation at a later date if needed. It's amazing how often a client will lose something that I have sent over to them and then claim that I didn't send it. By going back into the session archive file, I can show them that, yes indeed, that item was sent and on that day. A simple matter to cover your butt!

Try and get all of the paperwork and filing away into the archives finished as soon as possible after the session ends. I say this for two reasons: First, so that you remember everything and can log it correctly, and, second, so that you aren't faced with having to do a large number of sessions all at once because of the backlog piling up around your ears. That's never any fun, and it isn't how we like to use our time creatively. Better to get it done and out of the way as soon as possible.

Now, it's on to the edit and mix, and you know that your recorded performance is the best it was possible to get. To repeat, running a session is a major test of your multitasking abilities and your people skills. However, some things are just out of your control; I'd like to share of couple of experiences with you.

STUDIO WEIRDNESS

The studio can be a bizarre place to hang out and work in; you'll see some of the strangest things. Shouting matches, temper tantrums, and egos run amok are par for the course in the world of sessions. This is definitely not a part of the normal world that the majority of people know and live in. One day, we were recording a script for a training film for a pharmaceutical manufacturer. The producer was off site and directing the recording via telephone (not an uncommon procedure). The talent pronounced a chemical name without hesitation; the producer corrected him, "No, it's pronounced this way." "I beg to differ," replied our hero, "It's *this*." "Well, that may be, but at our company, we call it *this*." "Look," said Mr. Voice Talent, "I have a minor in chemistry from when I was in college, and I know for a fact that I'm right on this. If I'm going to pronounce the name of this chemical, I'm going to pronounce it the right way!" And away the two of them went, back and forth, and with the talent never understanding that it didn't make any difference—just say it the way the client wants it said and go to your next session. But no, tempers continued to escalate, and the voice talent ended up yanking off his headphones and slamming them into the control-room glass, shattering an otherwise good pair of phones. He stormed out of the studio. I picked up the phone and said, as calmly as I could, considering I had just lost an expensive pair of headphones, "Um, he just stepped out to get a cup of coffee . . . we'll be with you in a minute." I went into the lobby, and there was no one there. "Where did he go?" "I don't have the slightest idea," says the receptionist, "He just put on his coat and left." Which left me looking at the telephone, trying to come up with what I could tell the client. Mr. Ego had struck again.

One time, we were recording a math textbook (ah yes, show biz is glamorous, isn't it?) in a studio that had a "blind" booth. That is, you couldn't see into the booth from the control room because it was down the hallway, which normally didn't present much of a problem, although I prefer to have line of sight into the booth. The talent was seated at a table, and we were on our third day of recording the entire book—examples, equations, and all. Sometimes, it took a while to figure out how to describe an equation, and there was a time when the talent quit speaking and was silent for well over a minute. "Have you got it figured out, Charlie?" "Have what figured out?" "The equation," I said. "What are you talking about? I was reading along and all of a sudden you interrupt me. Why did you do that?" Son of a gun— he had read himself to sleep! Dozed right off in the middle of a sentence and didn't even know it. Time for a break.

I've done a lot of recording of campaign advertising for national political figures, and I had one client who was notorious for agonizing over every syllable of every word of one of his scripts. He was not only the producer, but the writer as well, and would often revise the script many times during the course of the session. The talent and I were used to it and usually didn't mind, because, while he sweated over his masterworks, the clock was running, and we were both billing by the hour. For one session, we were recording a 30-second radio ad; we had started at

1:00 P.M., and it was now 6:30 P.M.. Around and around the producer went with his copy, changing first this word and then that one. He suddenly looked at his watch, jumped up, and announced he was going to the Bruce Springsteen concert and would be back around 11:30 P.M. or midnight. We were to wait for him to come back, and then we would finish up. The talent walked out of the booth, put on his coat, and calmly said, "I quit." "What? You can't quit, we're in the middle of a spot!", replied the producer. "Tom, what do you think? Tell him he can't quit." "Well," I said, "just turn out the lights on your way out the door." Realizing that he had a full-blown mutiny on his hands, he asked for a playback of the last take that we had recorded (about a half-hour previously). "Yep, sounds perfect—have it on the air by seven tomorrow morning." Yikes.

I've had people throw full cups of coffee at one another, one time a laptop sailed across the room during an argument (luckily missing the intended victim), and various threats have been made. There have been practical jokes too numerous to name. (One of my all-time favorites involves a documentary project that we did. We had been working on it for over a week, and working with a producer from out of town who drove everyone to distraction by virtue of his pompous, obnoxious manner. This included the local director, who was also a very good visual artist as well. The client went to lunch (nothing for us, of course—typical for this guy), and the director pulled out a piece of lead foil film bag of the type used to ship film, so that the X-rays don't expose the film, and calmly traced the outline of an automatic handgun on it, cut it out with a razor blade, open the client's briefcase, and glued the outline under the lining of the briefcase. He closed the lid, and not a word was said. The client came back from lunch, picked up the briefcase and other belongings, and left for the airport. We could only imagine the scene as the airport security folks saw those X-rays and began the search for the invisible gun. Hope the producer caught his plane.)

All this and more are everyday occurrences in the studio. I don't believe I've ever laughed so hard or felt so inspired, while at the same time making some of the most lasting and deepest friendships, as I have by spending my life in the studio. Your job is to keep things moving along smoothly and not get caught up in all of the craziness breaking out around you. Have fun, but realize that your job is to give the client the best interpretation of their vision that you possibly can. Don't let anything interrupt your concentration on the job at hand. Above all, the vibe is king!

As the poet Rudyard Kipling wrote,

> *If you can keep your head when all about you*
> *Are losing theirs and blaming it on you,*
> *If you can trust yourself when all men doubt you,*
> *But make allowance for their doubting too;*
> *If you can wait and not be tired by waiting,*
> *Or being lied about, don't deal in lies,*
> *Or being hated, don't give way to hating,*
> *And yet don't look too good, nor talk too wise.*

Your Personal Recording Space

8

As we move more deeply into the twenty-first century, it is becoming more and more common for individuals, both engineers and voice talent, to have their own spaces for recording and production. However, if we could peer back through the dark mists of time, we could see a time when the only way to record and produce voiceover was to travel to a commercially available recording studio, there to be surrounded by incredibly complex and expensive equipment and gaze at the lord of his domain—the recording engineer. Master of the mysterious buttons and knobs, basking in the gentle glow of lights from the massive recording console, and performing tasks at once miraculous and yet comforting: Massaging raw voice tracks into a perfect final product. Now fast-forward to the modern age, where anyone with a basic grasp of how a word processor operates can far surpass our Merlin of that previous age. A long time ago, I began to notice more and more voiceover professionals asking my advice on various technical aspects of recording and editing. It slowly dawned on me that this early group of pioneers was seriously considering setting up their own studios! My initial reaction, naturally, was to withhold as much information as possible, while still remaining on speaking terms with them; after all, if their ambitions were successful, they were going to put me out of business. But, the longer I considered it, the more I came to realize that they were going ahead with their plans, with or without my help. It was a matter of either helping them or fighting them (a losing proposition, considering the number of actors seeking free advice). Well, I'm still in business and happily coexisting with the voice actors, the majority of whom now record in their homes in their personal recording spaces.

This chapter will give both actors and engineers an overview of what should be considered in taking the steps toward joining those who are already spending their time recording and producing from home, instead of facing the long commutes to run all over town for auditions and sessions. Always keep in mind that everything that has been said previously in this book still applies, whether that be relating to how acoustics work, or microphone choice and operation, or the importance of proper documentation and archiving of materials. The chapter is organized into four main sections, which should break the process down and remove some of the mystery about what has to be done to get up and running in your own home studio:

- The Space
- Hardware
- Software
- Delivery.

But first, some basic advice and suggestions.

THE BASICS

I am a professional studio engineer; as I've stressed throughout this book, I don't think the talent–producer–engineer team can be topped when it comes to turning out the best product, no matter how appealing working in your bathrobe might seem. Many people would like the convenience of having an in-house recording space—you don't have to travel to a commercial studio and you can be located virtually anywhere in the world where there is a fast Internet connection. Scripts can be emailed to you, and you can communicate with a producer or director located in a remote location, such as his or her office or car, in a distant city. You can deliver your completed work to the producer via the same Internet that delivered the scripts to you. Does this mean that you have to be an accomplished recording engineer? No, but you do have to have a very good understanding of the technical tools you will be using, especially your recording software of choice and computer operating systems. But my guess is that you already have a pretty good grasp of how a computer operates and how to create, name, and save files.

A word of caution for the professional engineer: By all means, help your friends out with the technical aspects if they ask. But remember that, when their personal studio is up and running, you won't see them much anymore. You'll be losing business! Consider a consulting fee to help them set up their space. I think that this is a fair arrangement, and the voice talent shouldn't have a problem with the idea. You can also suggest that they use your engineering services if they have a project that is complex or will need extensive editing and mixing. And, of course, more and more engineers are putting in their own home spaces, also. So, those among you of the engineering persuasion might want to pay attention to this chapter and think about some of the ideas presented.

Before we dive too deeply into this subject, I have to take a minute to remind you that, for the most part, this book deals with only recording the human voice and doesn't get into what happens in the control room (editing and mixing). However, for this chapter, I think that we have to address some of the basics, as the voice talent setting up her own space must perform those tasks as well as simply recording the narration. So, we will be dipping our toes into those subjects, although not in great depth. Also, keep in mind that I am not recommending specific pieces of gear in this chapter; those that I do mention are for illustration only. Each of us has our own preferences, our own likes and dislikes—what works for me and is a piece of "must have" equipment may not appeal to you at all. That's all right—and don't be intimidated by it. Never let anyone say to you, "Oh, you shouldn't be using that." If you like it, and it works for you, fine (as long as it sounds good and delivers the goods). My favorite loudspeakers may sound awful to you, but somehow I keep cranking out good-sounding work on them. Difference is what makes the world go around—and keeps the equipment manufacturers in business! Do your research; listen; evaluate; ask questions; try things out. What works for you, works for you. These are the types of question that the voice actors were asking me all those years ago—"Tell me why you're using that particular thingy and how does it work?" So, having said that, here we go.

Building your own space is not for beginners—you need a steady income to support it, unless you consider this a hobby and not a career. If you set up a personal studio, you are now a studio business, and you have to understand the tax implications, legal issues, and liabilities. Also, you will have to do marketing to get your voice and your projects out there. Remember, the goal is still a high-level, professional recording, and to get that is going to cost some money. Keep in mind that upgrades will be needed over time (software, etc.), but much of your investment will be a one-time, up-front expense (mics, preamps, acoustics, etc.), and so you don't want to shortchange yourself on these. Remember to take into consideration the hidden costs—cabling, lighting, maintenance, and so on. More on that a little later. When you go to a commercial studio, part of the studio fee goes to legal advice, marketing, maintaining equipment and new purchases, accounting, staff salaries, printers, telephones, and so on—the infrastructure. You will now be responsible for these additional office expenses as part of your home-studio business, so plan accordingly. And be aware that you are going to have to put in the time to understand the technological aspects of recording; this is what an engineer does, and you are now the engineer, besides being an actor.

I haven't made any mention of voice acting technique, developing characters, working the microphone, and so on; these things you already know by the time you get to this point in your career as a voice professional. Always keep in mind that this is a big step, and that you will now be running a business. Have a well-thought-out plan for keeping track of your time, as well as your invoicing and the monies received, and be aware of the tax liabilities that come with dedicating a portion of your home to your business. But today, it's almost a necessity if you are going to be successful as a voice actor or independent engineer, and the benefits of working in your own custom-designed space and in the comfort of your own home far outweigh the downsides of running a home business. Being able to record to your heart's content, without worrying about the constraints of a commercial studio's hourly rate, and being able to record at any hour, day or night, is a huge benefit to you and your product. OK, so you've decided to take the step and set up your own personal recording space; now, where do we begin?

THE SPACE

It's possible to record anywhere—and I do mean anywhere. However, the results that you will get are going to vary drastically. Sure, I could just set up a mic in my living room and begin recording my voice; the problem is that it will sound like I recorded it in my living room! Outside traffic sounds, airplanes overhead, the furnace and refrigerator running, and the neighbor's barking dog are all going to make their way into the recording. Not to mention the reverberation found in most living spaces. You're going to be after that tight, clean, "studio" sound. At this point, refer back to Chapter 3, Acoustics. Most home environments are reverberant and allow far too much noise from the outside to leak into our recordings, and these are the things that we try to control. In a pinch, you could try to put the mic in a closet filled with clothes—those make excellent absorbers. The problem with this solution is that it will sound like you're in a closet filled with clothes! And the lighting probably isn't ideal, and where do you stand? So, let's get a bit more refined and look at some options. Basically, you have three choices (well, actually four, if you consider either standing in a closet or recording in your living room to be acceptable):

1. Modify an existing space in your home or office.
2. Construct a custom recording room.
3. Purchase a prefabricated voice booth.

So, which to go with? What are the pros and cons of these three choices? By the way, option 4, do nothing and hope it all turns out sounding good, we won't consider for the professional voice actor or engineer. Do your research, and then do some more. Before buying a prefab booth or beginning construction, explore what other options you have in your existing space and be very clear in your end goals.

If you want to modify an existing space in your home, you can expect this solution to be the least expensive. Materials are relatively affordable and offer the advantage of being portable, which means that you can take them with you if you move. This solution probably won't permanently alter your home, which can be an attractive benefit if you ever sell the house or if you are renting. The setup and construction are usually easily done (with a little understanding of acoustics and what you're trying to achieve), and, by rearranging the wall- and floor-mounted treatments, you can change the acoustical signature of the room.

You could design and custom build a recording area in your home. This is probably the most expensive way to go. The upside is that you can design *exactly* the space that you desire and that suits your needs, with no compromise, except for your budget. The downside is that you will be permanently altering the space, and you'll in all likelihood have to either demolish your booth or leave it behind if you move. And, of course, landlords really don't appreciate this option. Building an isolated room within a room is not for the faint-hearted, and you'll have a lot of research to look forward to, in terms of both acoustics and building techniques. This is not a project for the weekend hammer-swinger—most likely, you'll need professional builders for the construction. And the problem with professional builders is that they usually don't have an understanding of the specialized needs of a recording space, and your finished room may be compromised.

This brings us to our third choice—purchasing a prefabricated voice booth. Usually less expensive than full-scale construction and modification of your home, but probably a bit more than modifying your existing space, this is the route that most home-recording folks travel. As mentioned in previous chapters, there are a couple of well-known designers and builders of prefabricated booths (such as WhisperRoom, VocalBooth, and StudioBox—see the Resources section at the end of this book for web-site information) that offer a great balance of price, configuration, and options for the home voice studio. Mind you, these pre-made booths can still run to several thousand dollars, but they offer the advantage of good soundproofing, design by a professional, solid construction, and the ability to add most anything you would ever want through the myriad options that are offered. Although it may not be the easiest task in the world to break one of them down in the event of a move, it is possible, and these prefabs don't permanently alter the building in which they're placed.

Modifying an Existing Space

Setting up a recording area in your home or office needn't be a huge project, if you pay attention to a couple of basic principles. You'll remember that the two biggest enemies of a clean, strong voice track are noise and reverberation. When modifying your existing space for recording use, reverberation is the easiest to control. Noise, generated both from outside the space and from inside, is much more difficult to control. Unless you absolutely soundproof the room, your neighbor's lawnmower is likely to intrude. And soundproofing can only be done with options 2 or 3, not just by altering the acoustic signature of your room. So, when considering this option, start by choosing the quietest place in your home or office. It probably won't be the kitchen, nor is it likely to be a living room facing the street. If your space is suitably quiet to begin with (such as if you are lucky enough to live in the middle of nowhere, with no one else around for miles), you're halfway home. Now we can concentrate on tuning the room to get rid of unwanted acoustic artifacts. This needn't be as complicated as you might first think. If you recall, reverberation is the bouncing around of the soundwaves within the space, and also the microphone has a definite pickup pattern. If your mic is a cardioid or hyper-cardioid type, the majority of the sound that the mic is picking up will come from the front. Rather than treating the entire room, it is possible to enclose the mic in an acoustic space by placing two panels treated with acoustic foam in a V shape and the mic (and, therefore, you) facing in toward the middle of the V. If you're still hearing some bounce from the room, take a third panel and put a ceiling on your construction. (No, this doesn't have to be permanently attached: Just throw it up there. Now you can put the whole thing away when company comes to visit.) A carpet on the floor will help reduce any reverberation from that direction, and *voilà*, a perfectly suitable acoustic space for the cost of three pieces of acoustic foam, some plywood to mount them on, and that's it. Not bad.

The downside, of course, is that you can't move around much, being surrounded by all of these panels. So, assuming the room is quiet enough to begin with, we can start to expand our workable space. This will take some experimentation to find the optimum placement of the acoustical treatments, but this is how most professional control rooms are built. You want to get rid of as much reverberation as you can to avoid a "live" sound—you are after that tight, robust, professional sound. Start by treating the front wall (the one you're facing) and the side walls. Then address what is bouncing back to you from the rear. If you notice a boxy, bass-heavy sound in your test recording, you'll most likely want to treat the rear corners of the room with bass traps. A diffuser on the rear wall can make a huge difference in the sound of your recording, so keep that in mind. The ceiling is also a place that you will probably have to treat, at least directly in front of and above the mic. And, once again, please carpet the floor with a not-too-thin carpet or rug. It not only improves your sound, it's much more comfortable to stand on for extended periods, and it gives your cat a nice place to curl up and nap. Check out some of the hundreds of books and online resources for building your home recording studio for more information. Sorry, but I did mention that you're going to have to do some research?

Custom Designing a Studio

The first thing I want to mention is: Why do you think that studio designers cost so much? Yep, we're talking some serious money if you want to custom build a dedicated recording area in your home. To begin with, you want to make it soundproof, and that means isolating the space from outside sound. The only way to truly accomplish this is to build a room within a room and detach the space from the rest of the building structure. This is also known as "floating" the room. Besides eliminating noise from the outside, there is also noise being generated inside the home, as well. This is usually in the form of heating and air conditioning running, and, to get rid of the sound of air movement and the motors that move the air, specialized duct work is required. Possible? Yes. Affordable for the average person? Not so much. This option is going to take much more research and, in all likelihood, the use of a professional contractor and builders, and, as I mentioned above, they may not be knowledgeable about the intricacies of studio construction. Some things that you specify in your plans may not make sense to them, and, thus, they may assume that you simply don't know anything about construction. So, they might "help" you out by altering the design to match standard residential construction practices, and you are left with a substandard performing room after all your hard work.

The take-away from all of this is that, yes, custom designing a recording space in your home or office is possible; in fact, many musicians have their own private project studios that compare quite favorably with any commercial facility (and sometimes surpass the room performance of the commercial studio), but you have to have the space, time, and money to do this right. The vast majority of voiceover professionals don't need this type of space in which to record and produce their narrations.

Prefabricated Voice Booths

I've discussed this subject in Chapter 3, Acoustics, and so won't belabor the point here, but purchasing a prefabricated voice booth is, to me, the ideal compromise for the actor wanting to get into recording at home. As mentioned above, the price is reasonable, it doesn't permanently alter the building, and the quality of these booths can be outstanding. You can choose exactly the space that suits your needs in terms of size and comfort, and you end up with a room within a room for soundproofing purposes. By choosing to go with a prefabricated room, you end up with the best of both worlds for a modest investment, and it will serve you for years to come, just like your microphone collection. You can adjust the size of the booth to fit the space available to set it up in, and there are ports that allow you to run cabling in and out of the booth to control headphone volume, your microphone cables, AC power, and so on. In short, it is a recording studio in a box (well, on a pallet, but that's a minor point). If you regularly record more than one person at a time, there are models available in various sizes, and you can custom order something not found off the shelf by one of the manufacturers.

HARDWARE

Assuming you now have the problem of where to record solved, it's time to move on to *how* to record your voice. You're going to have to take a short trip into tech land at this point, and there's no better time than now to find out what you'll need to get pro-sounding results. The first and most obvious is a microphone (covered in detail in Chapter 4, Microphones). Because this topic has already been covered, let's move on to the next link in the chain.

You are going to be constructing a "signal path" for your voice to get from your mouth to the computer and then out to the speakers. A simplified signal chain looks like Figure 8.1.

FIGURE 8.1

Simplified Signal Chain

Your final recording is only going to sound as good as the worst-sounding link in this chain. So, if you have a wonderful mic, but your mic preamp is a piece of worthless junk, don't expect to come out sounding like James Earl Jones. You don't have to spend a fortune building your signal path, and the price of quality components is falling all the time, but you have to pay attention to what you choose to record through. As you can see, after the microphone, the signal first goes to a microphone preamplifier, which raises the electrical signal put out by the mic to a level that other equipment is able to work with. This is a critical component, and, like microphones, all mic preamps have their own distinctive sound. You can find mic preamps with built-in compression and equalization, which although useful, aren't strictly necessary for a good sounding preamp. *Do not* rely on the mic preamp that is inside your computer: as with all of the audio goodies packed inside the computer, you are going to get what you pay for, and, in this case, the mic preamp is worth about twelve cents. At best, they sound terrible. So find yourself a good, solid mic preamp that comes with good recommendations from industry pros who record voice and know what they're talking about. It's tempting to think that your computer has all the essential hardware for recording that you'll need already inside it, but those components that make up the signal path are just not of sufficient quality to deliver air-quality, professional-sounding recordings.

Microphone preamps come in mono, single-channel versions up through multiple microphone inputs. As with the options of compression and equalization, they usually come with 48-V phantom power for use with condenser microphones also. They may or not have a headphone jack on them, but that is usually redundant with other equipment you will be using. Choose the version that best fits your needs and evaluate its sound and features before making the purchase.

Next in line is the A/D converter. Once again, the sound is going to change (but not nearly as much as with a mic or a mic preamp) from one model to another, and the choices seem endless. Remember that a microphone is known as a "transducer"— something that changes one type of energy into another? Well, the converter (or A to D or ADC) does the same thing: It takes the electrical signal from the mic preamp and converts it into the language that a computer understands, 0s and 1s. You can't get into your computer without one. Yes, the computer already has an ADC built into it on its sound card—well, enough said about that. Go find a good-quality ADC in the price range that suits you. Be sure that the model you select has the sample rate and bit depth that will work best for you; in this day and age, that will probably be 96 kHz at 24 bits. As with the mic preamp, converters come in a range of options as to the number of inputs, and most of them also function as digital-to-analog converters (DAC), to get back out of the digital realm and into your speakers.

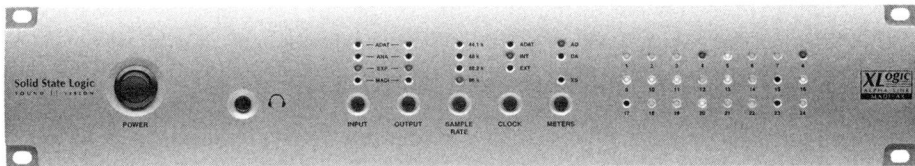

FIGURE 8.2

SSL AlphaLink SX Converter

Source: Courtesy of Solid State Logic

The converter connects with your computer via the USB ports, and magically you have now entered the digital domain. Whether you choose to work on a Mac or a PC is entirely up to you and the type of software that you prefer to run; both work equally well for audio projects, and there is no real difference between laptops and desktop computers. Assuming that your machine has the speed and storage capacity to handle high-quality audio, you should be fine. If you have a question about the machine you should have or should buy, check with a (knowledgeable) salesperson at the nearest computer outlet. You may want to take along a copy of the specifications for the software that you plan on running—it will list operating-system requirements, speed and storage specs, and more that can guide you in the right direction. Although dealing with digital audio takes a certain amount of horsepower, it's not as much as working in digital full-motion video or demanding graphics applications, and so the computer you choose doesn't have to be the highest of the high end. You just don't want it to bog down or crash when you're working with your audio.

There is one other piece of hardware that I cannot recommend highly enough: An external hard drive. The computer has a hard drive in it already—it could be an HDD (hard disk drive) or, if it's a newer model, perhaps an SSD (solid-state drive that works on flash memory). Some models have a combo drive that combines the

File Size	**1 Min.**	**10 Min.**	**30 Min.**	**60 Min.**
44.1 kHz/16	10.09	100.94	302.81	605.62
44.1 kHz/24	15.14	151.41	454.22	908.43
48.0 kHz/16	10.99	109.86	329.59	659.18
48.0 kHz/24	16.48	164.79	494.38	988.77
96.0 kHz/16	21.97	219.73	659.18	1,318.36
96.0 kHz/24	32.96	329.59	988.77	1,977.54
192.0 kHz/16	43.95	439.45	1,318.36	2,636.72
192.0 kHz/24	65.92	659.18	1,977.54	3,955.08

Table 8.1 Stereo Audio File Size (in megabytes)

best of both worlds. Owing to the phenomenon of old dogs learning new tricks, I still refer to any of these as the "hard drive"—sorry for any misconception. The purpose of the computer's internal drive is to handle the operating system and generate the graphic visuals that we see on the screen. Some of the internal drives are pretty big these days, but software requirements are constantly growing. I remember the first computer I worked on that had a 1-gigabyte drive—I laughed myself silly thinking that anyone would ever need that much space! But, like I said, the computer's internal drive is busy dealing with all of the computer housekeeping that allows us the speed and flexibility we demand from our machines, and it's best to have an external drive for your audio needs. As can be seen from Table 8.1, audio file sizes increase at an amazing rate as sample rate and bit depth are increased, and so the need for an external drive becomes apparent.

Keep in mind that the file size relates to the total amount of audio recorded, and not just the finished mix. It's likely that, for a 30-second spot, you will record much more than 10–15 minutes of material.

Also, when shopping for an external drive, be sure to get a drive with a fast access speed; 7,200 rpm is what you should be looking for. Any slower and your audio software is going to run either extremely slow or not at all. For this reason, thumb drives (those cute little USB sticks) cannot be used for recording and editing purposes. They're fine for quick storage and transportation of your sessions (assuming that the drive size is sufficient to store the size of your project), but you will not be able to actually work directly from the thumb drive. You'll have to download from the USB drive to your computer and into the audio software.

An intermediate step between the external hard drive and the thumb drive is the portable hard drive. Usually operating on flash memory (meaning no moving parts such as a spinning disk and read/write heads), the portable drives can be an inexpensive alternative to the stand-alone external drive. I recently purchased a portable drive with 500 gigabytes of storage, which is fast enough to run Pro Tools without a hiccup, for under $40, and there are terabyte portable drives available now for under $80. These little gems are great for moving your session materials from one computer

FIGURE 8.3

500-GB Portable Drive

Source: Photo by the author

to another and for storage of your archives. They are, for the most part, robust and trouble-free, offering good performance at a terrific price point. They're rugged, small enough to fit into a pocket or briefcase, and in many cases the perfect solution to your drive requirements. As a reminder, be sure to check the speed of the individual drive you are considering if you expect to be able to run recording software directly from it (rather than using it only for storage). Keep that 7,200 rpm number in mind.

We'll return to the subject of drives and storage in the last chapter of this book, "That's a Wrap," but, for now, know that the external drive you choose is essential to the smooth operation of your audio software and the confidence that comes with knowing that your work has been adequately protected and archived.

The Audio Interface

So, now we know that, from the mic, we have to go into a mic preamp, followed by an A/D converter, to get the audio signals into the computer so that we can work with them. Many audio pros take this modular approach, choosing each component for the particular and unique sound that the various makes and models give them. However, there is another way to consider, and one that is both convenient and affordable. This is the audio interface. The beauty of the audio interface is that it combines all of the functions that you're going to need in one unit; mic preamp, converter, USB connection to your computer, headphone amplifier and level control, and speaker connections to hook up your monitor speakers. Available from a single microphone input to eight or more inputs, you can choose the model that best suits your needs and requirements. As you will be recording yourself in your home studio, you probably won't need an interface with more than one or two mic inputs. However, consider your needs and what you might be recording; if you record music as well as voiceover, you may want a unit with more inputs.

The interface will contain everything that you need to record quality voice tracks—the mic preamp, the A/D converter, as well as a DAC and 48-V phantom power for your condenser microphone, and most often in a unit with a small footprint, so that it doesn't take up nearly as much desk space as individual units would. It's a perfect all-in-one solution to your voiceover recording needs, and many units come with audio software included (such as Avid's MBox line, which ships with Pro Tools Express software as part of the purchase price).

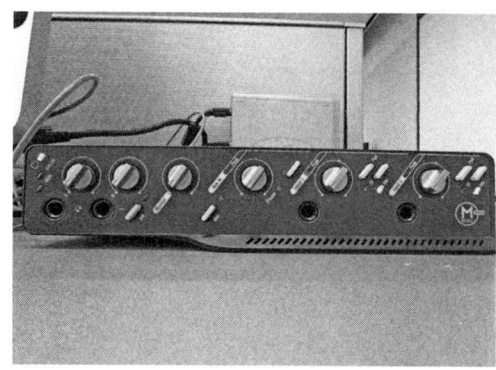

FIGURE 8.4

Avid MBox Pro (Front Panel)

Source: Tribeca Flashpoint Media AA; photo by the author

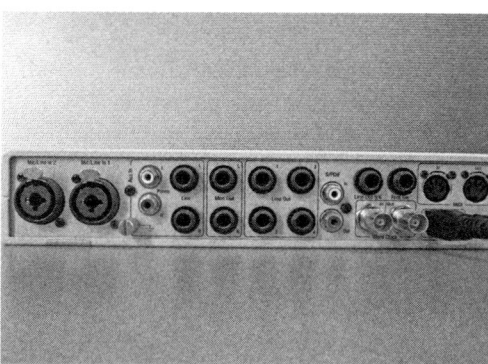

FIGURE 8.5

Avid MBox Pro (Rear Panel)

Source: Tribeca Flashpoint Media AA; photo by the author

Audio interfaces are available from a wide selection of manufacturers and come in various price points to fit most budgets, and most sound amazingly good, considering their cost. As with anything else when putting together your home studio, do your research, ask questions, and try out anything that you're considering purchasing before making your final decision.

SOFTWARE

As with anything else in the world, there are numerous choices when it comes to what software you are going to work with for your audio recordings. Known as "platforms," these various programs range from the free and extremely easy to use to quite costly and complex. In making your decision on which platform to go with, your main criterion is going to be the number of tracks that you are likely to need. Many programs are stereo programs, which limit you to working with only two tracks of audio at a time.

Easy to use, intuitive, and high quality, Audacity gives you good performance and some interesting features (and you can't beat the price). These types of program are great for recording, editing, and some very basic processing of your finished edits. However, if you are going to be adding music and/or sound effects and doing the mixing of all of the elements, you're going to need more than just two tracks, and, here, we move from the two-channel edit programs into multitrack platforms.

There are numerous software programs to choose from when considering a multitrack editor, but the main thing to keep in mind is that they all do basically the same thing; they just have different controls and terminology. Over the years, I have had experience with a wide range of software, and, if you work with a number of programs, you come to realize that, in the end, they're just boxes. Once you learn one, it's a fairly simple matter of learning another set of keystrokes, and moving to

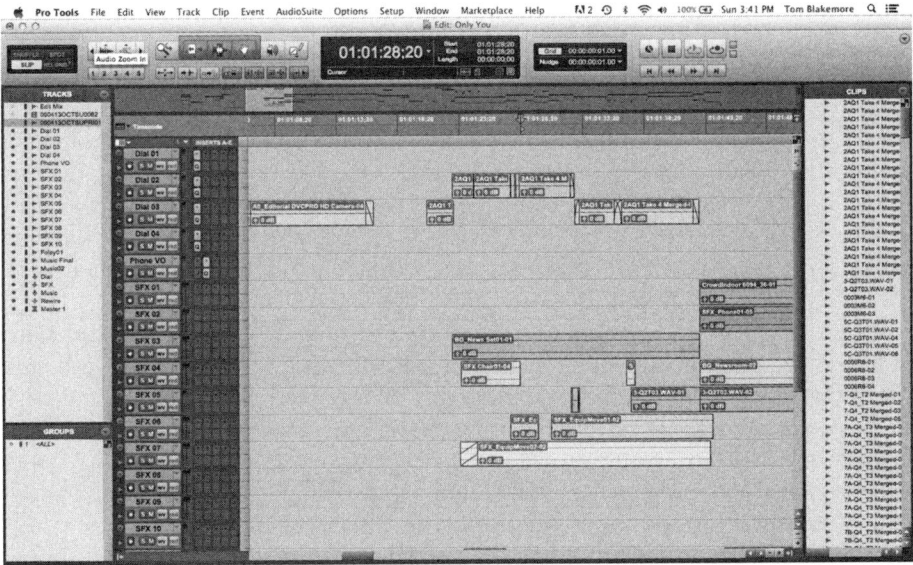

FIGURE 8.6

Pro Tools Edit Screen

Source: Photo by the author

FIGURE 8.7

Pro Tools Mix Screen

Source: Photo by the author

another isn't a huge deal. The main advantages of using a multitrack editor over a two-track editor are a larger feature set and access to a larger number of tracks for layering and mixing sound. The choices of processors (equalizers, compressors, reverbs, time compression and expansion, and so on) are greatly enhanced, and the world of third-party developers for these processors (also known as "plug-ins") is at your disposal. There are a large number of developers who put out plug-in packages, and the variety that these give you far exceeds the plug-ins that ship with the software you purchase. And, of course, the most obvious advantage to a multitrack editor is the large number of tracks available for you to work with.

Figure 8.6 gives an example of a Pro Tools session, and, besides what we see on the screen, you can add more tracks, name the tracks, perform volume, pan, and mute automation, and add place markers for editing reference points. Figure 8.7 shows another Pro Tools screen, this time showing a virtual mixing console.

With a virtual console, you can add and rearrange tracks, as well as doing complex signal routing, much as you would using a patch bay in the "real" world. Setting up sends, mixing stems, setting up headphone mixes, and so on, is a snap with a virtual console, as is controlling the levels of any given track by simply clicking on and dragging the fader up and down.

Many of the multitrack editors were developed with the music professional in mind and will have features (which you are paying for) that you will never need. In fact, nearly all of these software programs are loaded with all kinds of bells and whistles, of which you will probably only use a small handful. They are extremely powerful and deep; if you need the tracks, I suggest learning those parts of them that you need to work and ignore the rest. There may come a day when you'll need to dive back into the manual to learn how to perform some function or other, but, for the time being, just learn what you need to use on a daily basis.

Whether you choose two-track or multitrack software as your main audio production tool is up to you and your needs. It's easy to start off with something like Audacity, which is no-cost, and work your way up to more complex software as your needs expand and your digital audio chops improve. There's really no need to start off at the top of the line, especially if you are only going to be recording and performing some simple edits on your voice.

One other quick word on software before we move on: Besides your editing software, for your home studio you should have Microsoft Word. It's the industry standard word-processing software for business and is used by nearly all those producing documents such as scripts. Also, Microsoft's Excel is the standard spreadsheet software and should be included on your computer if you are running a home business. Both are available bundled in the Microsoft Office suite of software programs. For keeping track of invoices, money in and money out, and the like, I am partial to Intuit's Quicken line of data tracking for the economics side of things. With Quicken, I can set up specific codes for each client, each type of equipment that I purchase, and much, much more. Very simple to use and powerful, this is how I send my data to my accountant for tax purposes and keep track of how my business is doing at any point during the year (including how it stacks up against previous years and months). Recently, Google came out with a competing line of software programs that do these types of function, and Apple also has its own versions. No matter which you choose to go with, no home business should be running today without these types of software.

DELIVERY—THE REMOTE SESSION

It should go without saying that the main advantage of creating your own home studio is that you now don't have to travel all over town to do recording sessions. You can stay at home and let the world come to you. How does the world get there? Welcome to the telephone session. Many producers and directors love the convenience of directing the VO talent from the comfort of their office, not to mention the time savings of not having to travel to the studio. For them, it's a simple matter to pick up the telephone, dial the studio's number, and be connected to the narrator in the booth and the engineer at the console.

However, on our end, there's some technology that comes into play. Oh yes, you could just hold the telephone handset (or your cell phone) up to your ear, and the director would hear you reading the copy and then be able to give you direction for the next take, but that method is a bit clunky, to say the least. If you could do away with the phone altogether, simply speak into the microphone, and then hear the direction in your headphones as you would if they were in the studio with you, life would become much easier for all concerned. And, as an added benefit, your hands would be free as you spoke the lines and you could turn the pages of the script with ease.

To get to this point, we have some alternatives. The first is also the oldest, the easiest to set up, and the lowest-tech alternative. This is known as the telephone interface, or telephone hybrid. The hybrid is a lovely box that connects the telephone line to the console, allowing the director's voice to be fed into your headphones, and the console's output (the mic channel and the talkback) to be connected to the telephone, so that the director can hear you clearly. The process couldn't be easier: The telephone rings, you answer, and then push a button on the front of the hybrid. All of the interconnections between what is happening in the studio and what the producer is saying on the phone are instantly connected. Just like that!

The beauty of this type of system is that you can hear the director and she can hear you, as well as anything that you might play back during the session, such as the last take or an edit of previous takes. Most importantly, you are not recording their voice. It remains separated from what is being fed to the recorder (this is assuming that you set up the console correctly), and there is no feedback. One thing to obviously keep in mind is to keep the director's volume in your headphone down, so that you don't get bleed from the headphones into the mic, but you shouldn't have a problem. Telephone hybrids are convenient to use and affordable. No further technology is required (well, aside from the telephone and phone company's lines), and they are remarkably easy to use. There are a number of hybrids now available that work equally well with cell phones. Getting all of the connections made can be a bit of a challenge, but most hybrids come with good manuals, and the connection points on the unit are well labeled. The downside of using a telephone hybrid is that you are dealing with telephone-quality audio (not the best on a good day). Although fine for a reference and for the producer to judge the accurate delivery of the lines, the audio is not broadcast quality and can be affected by all of the usual audio issues of a standard telephone connection, such as static, clicks, and hums.

However, even though the director can hear you (we assume) clearly, and you receive her direction and make the adjustments in your delivery, there is one major problem with this system. You still have to figure out some way to get the material to the client. In the old days, this was by FedEx or a messenger delivery service. Today, of course, we can email files or place them in a dropbox system via the computer for the client to download. But, even with this relatively fast delivery method, there is still time involved at both ends. What would happen if you could eliminate that extra step of delivering the goods? And, to repeat, what is going down the phone line is not of broadcast quality.

The telephone hybrid works with the analog telephone line, which is limited in bandwidth, and thus the frequency and dynamic range are limited with this type of system. The unit acts as a transformer that changes the electrical signal of the telephone line and boosts it to line level for you to work with in your mixer, and then back again from line level to telephone level. Handy, yes, but not optimal sound quality. For this, we need something in the digital realm—the ISDN (*integrated services digital network*) line.

So, let's take a step up and get some real audio happening in this telephone session. Welcome to the world of ISDN. This marvelous technology allows you to transmit broadcast-quality audio in real time, solving the delivery delay problem and allowing the director to hear what you're reading as clearly as if they were sitting in the control room on the other side of the glass from you. But, as with anything else in the world, there are also some downsides. ISDN is the preferred delivery service today and has been for quite a while now, although IP solutions such as Source Connect are making inroads, even in professional studios. Many, many sessions are done this way, and the talent can be located virtually anywhere in the world. Your market is now the entire globe—and your competition is now anyone with a home studio and access to an ISDN line. Your competition is no longer the other voice actors in your town, but every voice actor anywhere in the world. For the voice talent today, it is becoming imperative to go this route.

However, before jumping into this, be aware of some basic facts. First of all, ISDN can be quite expensive. You are going to need a dedicated hardware unit (known as a codec, or coder/decoder) and the knowledge to connect the unit. You are going to have to have the telephone company install special, dedicated lines into your home, and the cost of the digital line required for ISDN is quite hefty. You'll need two in order to be able to transmit the bandwidth necessary for broadcast-quality audio. So, you'll be facing not only the installation charges from the phone company, but also a monthly fee for their use. Second, the person with whom you wish to do the session must have a matching unit; not any old ISDN unit will do. Because of the encoding and decoding into and out of the digital domain that must take place, the two units must be from the same manufacturer. Third, owing to the cost of using ISDN, most producers or directors will not have one lying around in their office, and so an additional studio charge is going to have to be paid to rent a studio with ISDN technology.

Alright, now that the downsides have been covered and it sounds like I'm trying to discourage you from getting into the world of ISDN, how does it work?

Once the installation is complete, from the front of the unit you use the numeric keypad to dial the number you wish to connect with. You'll notice from Figure 8.8 that there are two windows displayed on the unit; this is because you have to dial two numbers to make the connection (remember I mentioned you are going to need two digital phone lines?). The unit checks the connection, and, from this point, the operation is very similar to using a telephone hybrid. The output of your codec is fed to the console, so that you can hear the incoming direction from your producer, and your voice is fed from the console into the codec for transmission down the lines. Really, that's about all there is to it. The trick is in installing the phone lines (and being able to pay for them), purchasing the unit, and making all of the necessary connections. Once these things are taken care of, just push the "connect" button and start talking. At least, it's close to being that simple. Both the sending and receiving units have to be set to the same data settings, but that's pretty easy (there's a "select"

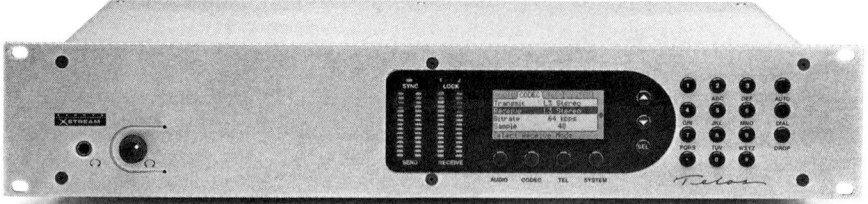

FIGURE 8.8

Telos Zephyr XS (Front Panel)

Source: Courtesy of Telos Systems/The Telos Alliance

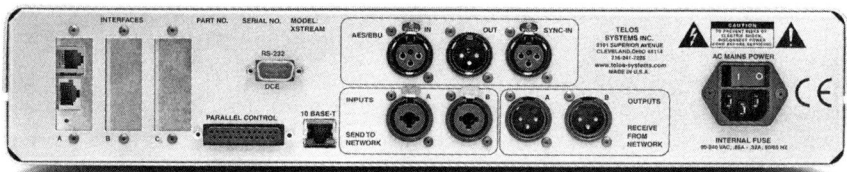

FIGURE 8.9

Telos Zephyr XS (Back Panel)

Source: Courtesy of Telos Systems/The Telos Alliance

button on the face of the unit to make your choice); if everything is set up right, it's fairly straightforward. You have now arrived at being able to send and receive full-bandwidth digital audio directly from your own studio. Using an ISDN isn't rocket science, but it is a major step to take in your home studio, because of the associated cost involved.

Is it possible that there is a desirable method of conducting a remote session that combines the ease of use and affordability of the telephone hybrid with the full-bandwidth digital audio capability of ISDN? The answer is: Kind of, at least as things stand now. Internet audio is constantly improving, and it won't be long before all of the telephone-based technologies are a thing of the distant past. When the Internet first started delivering audio, the audio was pretty terrible. Very low sampling rates and bit depth and excessive latency were the standard owing to very slow connection speeds and limited storage capacity. With today's high-speed broadband connections in most every home and office, things are changing rapidly. Internet-based solutions for the remote session are constantly getting better and only look to continue to improve. Software programs such as Source Connect are gaining traction and continue to gain regular users who, at one time, only had ISDN as a delivery method for their material.

With these types of program, there is still the limited drawback of both parties having to have the same software (just as with ISDN codecs), but the cost and ease of use of this software can't be beaten. Now, instead of having to have a dedicated piece of hardware to interface with the phone line and the audio console, your computer becomes the hub. Source Connect is a plug-in and works with Pro Tools and many other audio programs. There is also Audio TX; although more expensive than Source Connect, it is able to interface directly with ISDN and could be a nice bridging technology. Google is rapidly expanding its services into the media realm, and I fully expect that it will enter this arena soon, as well. Internet and VoIP technologies are constantly improving, as is the Internet's ability to stream artifact- and glitch-free audio, so this method of conducting the remote session is only to grow in the future.

ADDITIONAL COSTS

Now that you have made the decision to set up your personal home studio and you think you have the budget all figured out, you should take a careful look at hidden or additional costs that you will incur in making this happen. As I mentioned at the beginning of this chapter, I don't want to discourage you from taking this step, but you should be aware of the realities of what you're about to embark on. You've done your research and have a good handle on how much it's going to cost for whichever of the three options you decide is best for you and your space; you've chosen your mics, mic preamp, A/D converter and mixer. You're about all set, right? Let me tell you about a studio owner I worked for a while back: If one of the engineers came to him and said, "We really should have this thing—all the clients are asking for it, it will make our work more efficient, and it's only $700," his common reply would be, "How much is this $700 going to cost me?" His point was that there are always hidden costs that should be taken into account. Cabling, paying someone to integrate the new equipment into the patch bays, specialized mounting, and so on, all add to the cost of purchasing whatever you want to add. So, in thinking about your personal studio, don't forget to add in the cost of mic cables, USB cables, mic stands, copy stands, lighting, a good printer (for scripts and invoicing, as well as a myriad of other things), pop filter, headphones, possibly a headphone extension cable, and on and on. You probably won't think of everything until you actually begin setting your studio up, but think long and hard about what else you're going to need, beyond the obvious basics. All of these additional pieces can add up pretty quickly and add significant cost to your project. Don't scrimp on the quality of all of these pieces, either; you're going to have to live with many of them for a long time, and you don't want them to be constantly needing repair, nor do you want to buy the same thing once a year, when the old one wears out. As I said earlier, do your research, and then do some more. Ask established engineers or other voice talent who have already taken this step what you should take into account and listen to their answers.

One other key element of setting up your personal studio is going to be your choice of monitor speakers. What? I can't just do all of my editing and mixing on the same headphones that I used to record? No, no, no. Editing and mixing on headphones will not allow you to judge the finished product accurately, and wearing headphones for extended periods is just plain uncomfortable. Like everything else that goes into your studio setup, consider monitor speakers carefully and listen to as many different models as you can to judge what's right for you and your room. Rule number one is: You are not going to use the same speakers that are currently hooked up to your stereo system! Speakers made for your home stereo or surround-sound system are designed for that purpose and are optimized to give good results in a living-room type of environment. They usually are "colored" to optimize their sound, and, for a stereo setup, this is what you want. However, studio speakers, known as "monitor speakers," are designed to deliver a flat frequency response and should have no discernible "signature sound" of their own. Of course, all monitors have their own personality, which is why I prefer one set of monitors, and another engineer likes another. Some, I find, induce ear fatigue rather quickly, whereas, with others, I can listen for extended periods at relatively loud levels and never tire. Because each monitor speaker has its own personality, you will very often find multiple sets of speakers in a professional control room, and this gives the engineer a more accurate view of the sound he is creating.

When selecting your monitors, the first thing to remember is that the cute little speakers that came with your computer are worthless. Like pretty much everything else that comes included with a computer that is used for sound, these speakers just don't have a high enough level of quality for you to accurately judge the sound that you're working with. So, you are going to have to move into the world of professional monitor speakers. The monitors come in two basic flavors—passive and active. Passive speakers require a separate power amplifier to deliver the power that runs them, whereas active monitors have the amplifier built into the speaker cabinet itself. This increases the price of the monitor, obviously, but, on the other hand, you're not purchasing additional amps. Manufacturers love the active design because they can optimize the amp for the individual speaker design, and, thus, you end up with a better sound (at least in theory). Which direction you decide to go in is up to you, and either passive or active can deliver high-quality, accurate sound. Oh, and just to complicate matters a bit more, the sound of a monitor speaker can change drastically with the room in which it's placed or according to where in the room you set it up. Once again, try as many monitor pairs as you can get your hands on and critically compare the results.

Once you set up your personal studio, chances are you'll turn into a gear head (just like the rest of us audio geeks). There's always more equipment out there that you just have to have, that you can't live without. You'll find things that would be nice to have or that are just plain cool. When it comes to equipment purchases, I always stop and ask myself, "How will this make me money and pay for itself?" Sure, not every single thing you buy has to fit into this question—like I said, some stuff is just cool to have—but the majority of what you decide to purchase should pass this scrutiny. Otherwise, you'll be supporting your studio instead of the studio supporting and working for you.

THE MATTER OF TRUST

If you recall, we had quite a conversation in an earlier chapter on the matter of building and keeping trust. This is a vital aspect of your business and should always be kept in mind. When you get into setting up your own studio and begin recording from home, you are taking on the responsibility of running the entire session yourself and guaranteeing that you will deliver optimal audio to the client. When you go to a commercial studio, you are dependent on its mic choices, and those choices may not be the best for your individual voice. With your home studio, *you* are the ultimate arbiter of what mic you are going to use. You can custom choose your own mic and signal chain that best show off the true quality of your voice. The same goes for all of the other pieces of equipment that go into your studio. The client is counting on you to deliver the highest-quality final product that there is, on time and on budget. You are responsible for the quality of your recording, so give plenty of thought to the signal chain: Mic, preamp, interface, and storage. How much are you willing to spend?

You will be the actor, the engineer, and the producer's ears in your session, and you are going to have to get all of the roles correct. If you're technologically challenged and don't know a sample rate from a Studebaker, you may want to rethink your home-studio dreams. You are responsible for recording, editing, mixing, and delivering air-quality audio to the client in the format that matches their delivery specifications. This is all your responsibility, and all it takes is messing it up once or twice and that will be the end of that client. They will no longer trust that you are going to deliver what you promise to deliver, and they will find someone else who *can* give them what they expect from a professional recording studio. As I've mentioned previously, at the end of the day, all you have is your good name, and you should do all you can to protect it, so that you are known as someone who can be relied on to deliver the goods and who can be trusted.

Recording for Commercials

9

For sheer volume of voiceover recording being done, perhaps no other single category takes as much of our time as commercials. (I say this without any hard statistical evidence to back me up, but it has been my experience in the markets that I have worked in that commercial recording far outpaces everything else that we do.) It is not unusual for an advertising campaign to produce thirty, forty, or more individual commercials in a package. Obviously, this volume of work demands that we have a firm grasp of the intention of the campaign and a well thought-out game plan going into the recording sessions.

Many times, we may have three or four "base" commercials, and each of those will have variations in location and date, or other particulars that change, according to where the commercial will be broadcast or when a special sale is being announced. If we have a grasp of the intention of the commercial producers before beginning the recording, all of these individual variations can be planned and executed in an orderly, logical way. In this chapter, we will examine how to implement the process to arrive at an effective end product in a reasonable amount of time. Keeping the session(s) running smoothly and on time is one of your major responsibilities, and hopefully this chapter will give you an idea of some ways to do that. I'll return to this subject at the end of the chapter, but now is a good time to remind you that doing commercial recording is a great way to sharpen all of your skills in the studio, whether you are reading or engineering. Just the sheer repetition of doing this work makes you faster and sharper.

Note: This book concerns itself with the act of recording voiceovers and doesn't get into the editing and mixing processes much. There are some very good tutorials on these subjects, both in book form and online, that are easily found, and these skills are outside the scope of this text. However, in discussing specific recording situations, some mention of editing, processing, and mixing must be made, and in this chapter I will give you some specific ideas on these matters.

Recording commercials on a regular basis is great training to become very fast and efficient in everything you do. Your studio setup time becomes faster, and so do your editing, mixing, and overall decision-making. What I've learned in years of commercial recording pays great dividends in working in film and television. This is also true in the music world, by the way. Because advertising music sessions operate at the same speed and intensity as the voice-recording sessions, music engineers become used to the short setup time and tracking that they have to do. There is simply no time to take two days to get the perfect drum sound; the engineer must know exactly what he is going for and know how to achieve it in a very short amount of time, live with the results, and move along tracking other instruments. This translates into a really sharp engineer when it comes time to do album work. In the voice world, the same skills apply: Choose your microphone, set it up, record, edit, and go to the next session. You'll be amazed at the depth of your experience in a short amount of time—*if* you are attentive to what you are doing and work at it.

We have more than a hundred years of experience with advertising theory that has been developed, and, if you learn a little about this, you'll begin to understand why things are done the way that they are. What initially may look arbitrary and/or counterintuitive becomes a bit clearer, and you'll come to see that there is probably a very well-thought-out reason for some of the decisions that are made. The basis of all advertising theory comes down to "problem/solution." That is, in all likelihood, the commercial will state a problem—"Say, are your clothes looking dingy and gray?"—and a solution—"Well then, try new and improved Sudso." Pretty basic stuff, but this is how most commercials are constructed. There are also legal implications to the advertisement, mainly relating to meeting various federal rulings on truth in advertising (that is, you cannot claim something that is not demonstrably true, such as saying that this product is far superior in taste to the competitor) and on clearly stating various restrictions, side effects, annual percentage rates, and many other things. Knowing all of this will go a long way in clearing up any confusion you may have about the structure and intent of the advertising.

Commercials can be produced for local, regional, national, or international use. Your approach to recording and how you do it doesn't change, but the stress level does; there is much more money at stake for airtime, union talent, music licensing fees, and so on, as we reach a broader audience. There are also varying pay categories, for international usage, for what is termed "all-use" (including Internet) that encompasses any and all uses of the commercial, and for a one-time fee to use the material in perpetuity (what is known as a "buy-out"). Becoming familiar with these categories and their associated costs should be a part of your education in the field of commercial recording. In the United States, commercial recording is governed by union agreements between the VO artists and the producers of the commercials: there are pay schedules that have been bargained for between the producers and the members of the SAG/AFTRA union (a merger of the Screen Actors' Guild and the American Federation of Television and Radio Artists). Part of the agreement is the paying of royalties (also known as "residuals") to the actor over a specified period of time, for every time the commercial (the actor's work) is broadcast. Unfortunately, recording engineers are not covered by union agreements of this sort, and so we don't receive any ongoing remuneration for our work, after the session is completed.

Apart from radio and television commercials that advertise a product or service, there is a category of spots known as public service announcements, or PSAs. All of the same strictures apply that we know from commercial announcements. The PSA informs the public about any number of issues, from warnings against texting while driving to reminders to set your clocks ahead at the next time change for Daylight Saving Time. The cost for airtime and the talent rates are usually discounted for the PSA spot, but the all-important time factor still comes into play.

THE DICTATORSHIP OF THE CLOCK

Commercials are also known as "spots," short for "spot announcement," and I will use both "commercials" and "spots" interchangeably throughout this discussion. The major difference between recording for most other forms of medium and recording spots is, of course, the commercial has a time limit that must be adhered to. Commercials most typically are recorded in 15-, 30-, or 60-second increments, and this is a function of the airtime that broadcasters make available for advertising messages during their programming. There are 10- or 120-second blocks available, but they are not as common as the 15-, 30-, or 60-second spots. Regardless of the time limit of the individual commercial that you are recording, these times must be rigorously adhered to. One-tenth of a second too long, and your finished work will be cut off on the air, and too short is equally bad. It all comes down to broadcasting schedules: If a number of spots run short, by the end of an hour, the broadcaster has time on its hands that it hadn't planned on, and no way to fill it. With a live host (a DJ, for instance), the host can ad-lib and fill the time, but, with television programs where everything is prerecorded (both the program and the spots, as well as promos, bumpers, and so on), there is no way to kill this extra time except for dead air (that is, nothing being broadcast). And dead air is the broadcaster's worst nightmare.

As I mentioned, keeping the session running smoothly and on time is one of your responsibilities, and this is often a struggle. I've had six commercial sessions booked back to back in a day, and, if one runs long, everything in the schedule starts to slip behind. You have to learn how to say, "Sorry, you're out of time—we'll have to pick this back up again in three hours." There are a couple of problems if this situation happens: The airtime may have already been bought, and what you are recording has to be on the air immediately. Second, if the voice talent has to return to the studio at a later time, he may now cost more than was budgeted, or the actor may not be available at the later time. So keep things moving! This is where your experience comes into play—you have to keep in mind how long the editing, sound-effect selection, and mixing are going to take and factor that into the session schedule. And, once all that is completed, you will still have to prepare materials for delivery to either the producer or to the broadcasting stations.

You can see that we have two problems with commercial recording: First, the spot has to be the *exact* length; and second, the individual recording session must not go over its scheduled allotted time. We'll take a look at some strategies to get the spot to proper timing; session timing is up to you. Because time is of the essence, have a game plan going into the session. As mentioned in Chapter 5, you should have already taken a look at the script(s) and timed yourself reading it. You probably have an idea of what sound effects you'll need and what style of music you might have to find, if the commercial isn't using a prerecorded piece of music. Of course, you should have a game plan for any session, but, because commercial sessions can be quite short, you'll have to have a good mental roadmap of where you are going before you begin the session.

As for your session management, my only advice is to keep a close eye on the clock and know when to begin moving things along—remember that you will probably need time for more than just the recording (editing, SFX selection, mixing, duplicating, and so on). For each individual read of the script, it's essential to have a stopwatch in hand so that you can give an accurate reporting of the time of the read to the VO talent and the producer immediately. Don't slow the session down by having to stop and then measure the length of the read in your DAW program. By using a stopwatch, you can call out the timing in tenths of a second, and not frame counts; it's much easier for most people to understand, "that's three-tenths too long," as opposed to, "that's ten frames too long." An important skill to master is accurately timing the read; you can't be sloppy with your stopwatch and you have to start and stop it exactly with the talent's performance. If the actor takes an unusually long breath or pauses during the read, you should be prepared to stop and then start your clock accordingly. Most experienced voice actors, as well as producers and engineers, have a clock running in their head and can tell when 30 seconds is about to expire and make adjustments in the read.

An important consideration in choosing a stopwatch is that it doesn't make an electronic beep when activated. This is especially true if you are going to be the one recording your voice—the mic will pick up the sound of the beep, and you wouldn't want that in your recording. For this reason, many voice talents choose a watch with a sweep hand—that's right, the old-fashioned analog stopwatch. Smart phones almost always have a stopwatch function, either counting up or counting down, and the audio of the phone can be muted to silence the activation beep. I think that another thing to consider is for the watch to have a "split time" function. With this function, you can time individual sections of a read, as well as get an overall timing of the whole read. This can come in handy for recording television spots, which at times are to be added to an already edited picture. This would be the case if we were not watching the picture while we recorded the voiceover. In this case, instead of simply making sure that the read comes in at 30 seconds, we want to have each section of the read match an existing timing from the picture. So, sentence 1 might be 5.5 seconds, lines 2 and 3 together might be 8.7 seconds, and so on. To ensure that the timing of the read matches the picture, we need to know the timing of each of these sections, and this is where the split times come in handy.

An alternative to the hand-held stopwatch is the studio timer. Studio timers are manufactured by a number of companies, but most have the same features: Large numeric displays, the option to count up or count down, and a remote function, so that the person operating the clock doesn't have to be right next to it (at the recording console, for instance). The timer can be mounted on the studio wall, and the voice talent gets an immediate readout of how many seconds remain for her to finish the reading. Most studio timers do not have split-time functions.

HOW LONG IS TOO LONG?

If we have a 30-second commercial to record, how much of those 30 seconds should the voiceover be? Figure 9.1 shows a typical radio commercial script. You'll notice that it is broken into two parts: The portion for the narrator (marked NARR) and the portion marked TAG. Quite often, but not always, these two sections are voiced by two different people. So, the question becomes, if the total length of the spot is 30 seconds (or :30), how long does each of the sections have to be? And how do we arrive at our determination?

The most efficient way to record this situation is to record the tag first, and then the body of the copy. Because the body of the copy is longer, it will be easier to make adjustments to the timing of the read. Also, tags are often dictated by legal considerations, and the wording must not change, whereas the body copy can lose or gain a word or two, usually without getting the legal department involved. Also, it is much more efficient, during the session, to have each of the VO talent record their lines individually, instead of having both in the studio at once, with one actor waiting to record his 5 seconds of copy for the tag while the other actor struggles

WRENCHMASTER

"Summer Sale: :30 Radio

H19L5430R

SUMMER MUSIC UNDER
<u>NARR (:25):</u>

It's that time again for summer vacation with the family, trips to
Grandma's house, and afternoons at the water park. But is your car ready for
summer? Come in now for Wrenchmaster's Summer Sale and get a five-
point brake inspection for free, plus 20% off pads and rotors. And don't
forget a summer oil change – with a new filter for only $22.95.
Make sure that your summer travels are safe and problem free – only at
Wrenchmaster! It's summertime – *relax*!

<u>TAG (:05):</u>

Sale prices good June 15th through July 20th. Not good with other
offers or sales. Not valid in Arizona or Louisiana.

FIGURE 9.1

Radio Commercial Script

to read through the opening 25 seconds of the spot. Remember, union voice talent is expensive, and to pay someone to stand around to say just a few lines is not a good use of the budget. Also, the studio setup becomes faster owing to using only a single microphone instead of setting up two, two sets of headphones, two cue mixers, and so on. Once again, we're striving for maximum efficiency in the session.

There are times when a number of spots will be produced that are all identical except for one line that changes; for instance, "Sale starts Sunday/Monday/Tuesday/Wednesday," and so on. The best way to ensure that the spot comes in on time is as follows: It is recommended that you record the longest variation first (in the above example, record "Sale starts Saturday" first) and use that as your base spot to cut all the other variations into. In this example, the days of the week are all of the same relative length, but there will be times when the variable element is much longer or shorter in time. If we were to record a commercial that would include the names of the cities Boston and San Bernardino, you can see that the timing would change significantly, and you must make allowance for this. By recording the longest variable first, you can much better make sure that you are not squeezed for time. It's much easier to slow down on an insert line than to try to read it in an unreasonably short amount of time.

Many times, the tag will be a legal disclaimer—ads for insurance, auto companies, pharmaceuticals, investments, and the like will have legal copy that must be included with the commercial (annual interest rates, "Past performance is no guarantee of future earnings," warnings of possible side effects of medication, and so on, are mandated by the federal government, but take away precious time from the body of the spot (the "sell"), and so there are two approaches taken with this type of copy: Either have the actor reading the tag read it as fast as humanly possible, or have the engineer time compress the read to a point far faster than a person could possibly get the words out of their mouth. The producer of the spot realizes that this is legal wording that is required, but doesn't particularly care if the audience understands every word. Again, the goal is to get through this wording as quickly as possible, and how an individual producer deals with it will vary. However, the technique remains the same: Record and time the tag first, and that will determine how much time there is for the body of the commercial.

Commercials are often produced with music and sound effects, and these are usually noted on the script. As the engineer, you must determine how long these elements will play in the clear without having words spoken over them. Be sure to take this into account when timing the script and the actual reading during the session. If you are dealing with a 30-second script, and it calls for a piece of music to play continuously underneath it, try to leave a beat of music at the top of the spot in the clear, to establish the music, and a beat at the end as a "button," finishing off the spot. This might leave you :28.5 for the total read from the VO talent, and this must be determined prior to starting recording takes. Again, for a smooth and trouble-free session, preparation is the key, and getting these timings is part of that process.

This same technique is used for any sound effects that are called for during the course of the spot. How long should a particular effect be heard in the clear during the read? If there are both effects and music playing during the spot, there is now less time to speak the words. An experienced writer will take all of this into account when constructing the spot, but, as either the engineer or the talent, you may have to inform the producer that, by putting all the effects and music in, the finished spot may run excessively long (and remember, a tenth of a second is too long!). The more elaborate the effects, the more time it will take for them to play out. While reading the copy, the narrator can just take a beat at the places where effects or music are called for, and the timing of the pause can be adjusted later to match the length of the effect being inserted. Just time the read carefully, starting and stopping your timer as needed when the talent takes that beat.

Many commercials have a piece of music written for them that contains lyrics. Known as a "jingle," this music is intended to brand the product in a memorable way and keep it in the mind of the listener. The vocals can come at the head, in the middle, or at the end of the spot—it goes without mentioning that the length of the vocals must be taken into account when figuring out how much time the narrator has for the body copy. You can record the voice while this music track is playing in the performer's headphones to help with the timing; just be wary of headphone bleed from the music track leaking onto your voice recording.

What happens if the read consistently runs long and you can't possibly cut any words? There are a couple of ways of dealing with this situation, but they should be done with a light hand. While reading, it's obvious that the actor must breathe; no one can give a believable performance for 30 seconds on one breath of air. If your reads are consistently long, you could remove some of these breaths. However, if you remove all of them, the resulting performance stands the chance of sounding robotic and inhuman. I've found that a spot with all of the breaths removed makes listeners feel very uncomfortable, but they don't know why. Of course, if there is a music bed playing throughout the spot, it will hide the fact that you have taken out the breaths, which are much quieter than the words being spoken. The goal is to keep the finished production sounding natural, and so you can't remove breaths at random; there are natural places where we expect a person to breathe while speaking. Plan which and how many of the breaths you want (and need) to remove to bring the read in on time, and do this editing judiciously. Our goal is to keep the read sounding natural, and removing breaths can result in running the sentences together in a way in which a person could not possibly speak. Having too short a time between sentences sounds as bad as any other editing mistake, and this should be avoided. Keep thinking, "Could a person actually speak like this? Does this sound natural?" Consider removing only a portion of a breath in this case; you should know that if you cut off the beginning or the ending of the breath it will sound clipped off, so (carefully) remove the middle portion of the breath sound. A slight cross-fade usually helps to hide the edit, and you are beginning to shorten the read a few tenths of a second at a time. Keep this editing process in mind during the course of recording, and it will help you to understand how much time you can make up if the spot is running long with the read.

In the digital age in which we live, we have the ability to speed up or slow down the length of a reading without affecting the pitch of the voice. By using time-compression and time-expansion tools found in nearly all digital recording software products, this is an easy process: You can simply enter in the target time to which you would like to adjust the total time, and the software does the calculations. *Voilà*, a perfectly timed read. Not so fast. As with removing breaths, or with any editing that you do, naturalism is our goal, and it is easy to overprocess the voice using time compression/expansion. If the voice is overprocessed, digital artifacts in the form of "glitches" or "stuttering" begin to be heard, and this can't be allowed to happen in our finished production. You may have to do a judicious amount of time compression *and* breath removal to arrive at the natural sounding read that you're striving for.

One further word on timing the commercial reading: It may seem that the easiest solution to the problem of too many words for the time allotted for the spot would be to remove a few of the words. However, here's the thing: There are a number of very strict federal guidelines that govern what can and cannot be included in commercials. Therefore, the wording of the commercial nearly always must pass through the legal department of the advertising agency for approval of the language contained in the script. A large number of people, from the client to the copywriter, lawyers, and producer, have signed off on these words, which then are passed to you for the session. The laws governing truth in advertising, liability, and many other issues have been taken into consideration. If the copy is to be changed, it should be done very, very carefully, and you must *never* take it upon yourself to do so. What might seem to be an innocuous word change can have very large implications. There are often forces at work of which you are probably unaware, when it comes to working in commercial recording.

THE COMMERCIAL SESSION

For this and subsequent chapters in this book, refer back to Chapters 5–7 for specific tips on the recording process for voiceovers; here, I'll try and give you some information on commercial recording and some techniques that might help you with this recording situation.

Before the session begins, make sure you have some sharpened pencils, and have those take sheets handy. If you recall, you should have reviewed the script in advance if possible, and read the script aloud to make sure that the spot comes in on time. Reviewing the script will alert you to how many mics you should have set up and ready to go (along with sets of headphones and other equipment) and what types of sound effect and music you might need, and you will be ready to discuss any problem areas that you find with the producer. When the VO talent comes in, be sure to mark your copy of the script with any word changes that are made in the initial conversation between the actor and the producer. Of course, you will have plenty of (room-temperature) bottled water on hand, as well as anything else that you might anticipate the actor, producer, and any others sitting in on the session might require.

The first thing to determine is the number of actors that you will be recording at the same time. For a single narrator, the process is pretty straightforward. Just keep the actor on mic, provide a good cue send to her, and pay attention to documenting and logging of takes. With two (or more) actors, you have to make sure that the mics remain as separate as possible; if you are intercutting lines, you don't want bleed from one narrator getting into another actor's mic and lessening your chances of having a clean intercut. The best way to ensure a totally clean recording is to mute one microphone when another actor is speaking, or, at the very least, lower the fader for that mic significantly. However, in practice, this can be tricky, especially when the two actors have lines that come one after another very quickly, or when there is an intentional overlap of lines. It is often more effective to handle this in the editing process by muting one of the tracks when that person isn't speaking. Figure 9.3 shows a screenshot of this technique in a Pro Tools session.

In Figure 9.3, you can see that, by muting each track in turn, or "checkerboarding" the tracks, the chance of off-mic audio and possible phase problems is avoided, making it much easier to process and mix the two voices and achieve a consistent sound between the two. This technique proves useful in any recording situation with

FIGURE 9.2

Voice Waveforms. We can see that, when Narrator #1 is speaking, we still observe the waveform of that voice on Track 2—this is the off-mic bleed from Narrator #1 getting into the other actor's microphone

two or more voices, whether that be commercial recording, interviews, long-form narration, or any other. It is fast to perform and will greatly increase the quality of your finished production. A word of caution here, however: It may be tempting to skip this editing function by having a noise gate in the signal chain for each mic and attempting to automate this process. I've found that using a noise gate for this purpose is quite often audible (that is, you hear it pumping), and the chance of cutting off a soft beginning of a word (such as with an "h" or "f" sound) can become a real issue. It's much better to take a couple of minutes and checkerboard your tracks with editing. If you do make a mistake and cut off part of a word, you can easily restore the missing sound; if you use a noise gate during recording, you can't go back and grab the sound again. This is true with using any type of processing during recording: If using a digital workstation plug-in, you can always adjust after the fact if need be, but, if you're using outboard processing equipment, you can never go back again and change it. You have to think this through very carefully if you choose to use processing on the recorded tracks and be prepared to live with the consequences. I highly recommend recording everything "dry" and making the processing decisions in the editing and mixing stages of production.

What type of microphone should we choose for commercial recording? As pointed out in Chapter 4, Microphones, every individual is different, and the way a specific mic will react to a person's voice is going to vary. Let experience be your guide

FIGURE 9.3

Track Muting

when it comes to mic selection for the individual actor. You should also take into consideration the final playback form that your recording is going to have: Is this commercial intended for television? Radio (AM or FM)? Is it intended to be played in a movie theatre? Each of these has its own sonic requirements, and this may play a part in which mic you choose to set up. For instance, in most cases, and for general voice recording, I usually will put up a Neumann U87. However, if the producer informs me that this is going to be a television spot, and that it should cut through the other commercials on the air, I might instead go with an AKG414. Although this particular mic is far from my favorite for voice recording, owing to its unique frequency-response characteristics, it would be a good choice in this situation because of its clarity and ability to stand out above a more rounded-sounding mic such as the Neumann. For playback in a movie theatre, I could go with a Neumann U67, which has a warmer sound and, in this environment, doesn't have to compete with other noise and program content, such as would be found in a home viewing situation with television. Again, let experience be your guide—if you are unsure about your choice, set up two microphones side by side, and, after processing and mixing, the choice will start to become clearer to you. (Throughout this book, a number of the guest contributors weigh in with their thoughts on microphone selection, either for voice recording in general or for specific types of program material; consider what they have to say on the subject of mic choice and put it to use in your thinking.)

There may be times when you reach for an unusual microphone for a specific intended effect. If your character is supposed to sound like an old-time radio announcer and you can put your hands on a vintage RCA mic, you'll be way ahead of the game as far as getting an authentic sound. There may be instances when you will want to use a shotgun mic, especially the famous Sennheiser 416 (refer to some of the "Insight" comments from contributors) or a Schoeps shotgun. I've recorded some commercials using an old, beat-up ElectroVoice 666—your choice of mic all depends on your desired result. My advice is to keep experimenting and see where it takes you.

As with any type of recording, documentation becomes very important during the process, and this is especially true of commercial sessions. Owing to the large number of takes that may be required to arrive at the desired performance, keeping track of the takes is vital. A nicely designed take sheet, made specifically for recording spots, is of immense value later when it comes to the editing stage of production. Refer to Figure 7.2 for an example of a take sheet for commercials. You will note that there is an individual take number, total time of the read, and any comments that you may care to add to help in selecting the proper takes. Along with the take sheet, the recording itself should have an audibly recorded slate from the engineer. In a digital workstation environment, the talkback or slate microphone can be routed to a separate track, and, on playback, we can identify which take we are listening to by the take number that was announced during recording. In the days of analog tape, when the "Slate" button was activated on the console, it enabled a low-frequency tone (typically in the 40-Hz range) to be recorded along with the engineer's voice. On high-speed winding of the tape, the pitch of the tone was raised, and you could

FIGURE 9.4

Auto-numbering of Takes

hear the individual slates going by and count them by the sound of the "boop" that the tone created. With digital workstations, we can let the software work for us with a function known as "auto-numbering." Each time the workstation is put into record, the software amends the name of the recording with a number. In this way, the first recording might be called "VO-01" in the software, the next "VO-02," and so on. It now becomes a very easy task to identify and play back any take that you are looking for, without having to go back and forth searching for that take, thus taking extra time during the session. You now have both a visual reference to find the take you're looking for and an audible confirmation of the take number.

I found out about the importance of the voice slate in a strange manner one day. My old friend, voiceover actor Herb Graham, was in our studio doing a commercial session. At the end of recording, the producer asked if Herb could do a quick read of a spot for IBM (this was back in the days of IBM being a major player in the world of desktop computers). The producer wanted to play the read for IBM as an example of the type of campaign that the agency might produce to promote the sale and use of IBM computers. "Just read it down one time, Herb. No need to get too fancy with this, it's just a demo spot for them," says the producer. So Herb read it through once (I didn't even bother to time it—this was only for the purpose of demonstrating the type of spot that the agency was thinking of), and no one paid much attention to what we were doing. You know, down and dirty. Herb read the spot, I put it on an audiocassette (this was a while ago, wasn't it?), and the producer was off, as I continued to duplicate the real spots we had done to later send out to the radio stations. The next day, the producer called with a rather strange request. He wanted me to record voice slates from "Take 1" to "Take 7," then duplicate Herb's single read seven times, and put each voice slate before a copy of the read. Not knowing what was going on, I just did as he asked me to do and sent the results over to him. The following day, he came in all smiles and could hardly contain himself:

> *Well, we played it for IBM, and they really liked it but wanted to hear the other takes to see what options they might have in the delivery. I played them the seven copies you sent, each with their own slate number, and after a lot of deliberation, the IBM people decided that we had picked the right take, although they thought the beginning of take 4 was a little stronger and that take 6 was a bit too fast and thrown-away in the read.*

Ah, the power of suggestion! Incidentally, Herb's lone reading was produced as is and ended up as a national commercial that ran for over six months.

INSERT LINES AND REDOS

We've talked about the strategy for recording spots with tags and how to avoid timing issues in these spots. But what happens if you have to redo a line of copy at a later date? It not only has to fit into the original recording seamlessly soundwise, it also has to match the original timing. Consider thinking about this situation in terms of doing an ADR session. ADR is used in the film world to replace lines of dialogue recorded on location that cannot be used for one reason or another. Perhaps there was a noise issue, such as an airplane flying overhead during the performance. Or maybe the director wants to change the emphasis on a word or alter the emotion in the read. Sometimes, technical problems happen with the recording equipment. In these cases, the actor is brought into the studio, and the picture is projected in a loop of the line(s) to be rerecorded. The actor watches the film for timing and lip-sync, and the engineer and director are paying attention to performance while guiding the actor toward the performance that is desired. In the world of commercial recording, especially radio commercials, we generally don't use the picture for a timing guide, unless the actor is on camera, but instead play the original read in the narrator's headphones so that he can find the correct timing. However, there's just a little bit more to it than that.

Having to redo a line of copy in a commercial is more common than you might initially imagine. In fact, it's quite common. There can be any number of reasons for this: Something was mispronounced, the date of a special sale has changed, the spot might need to be customized for a different city, and so on. I've had instances where a spot has been revised many, many times over the course of a couple of years or more, and the goal is to always make the new insert invisible for our listeners, while keeping the original timing of the spot. If you run into this

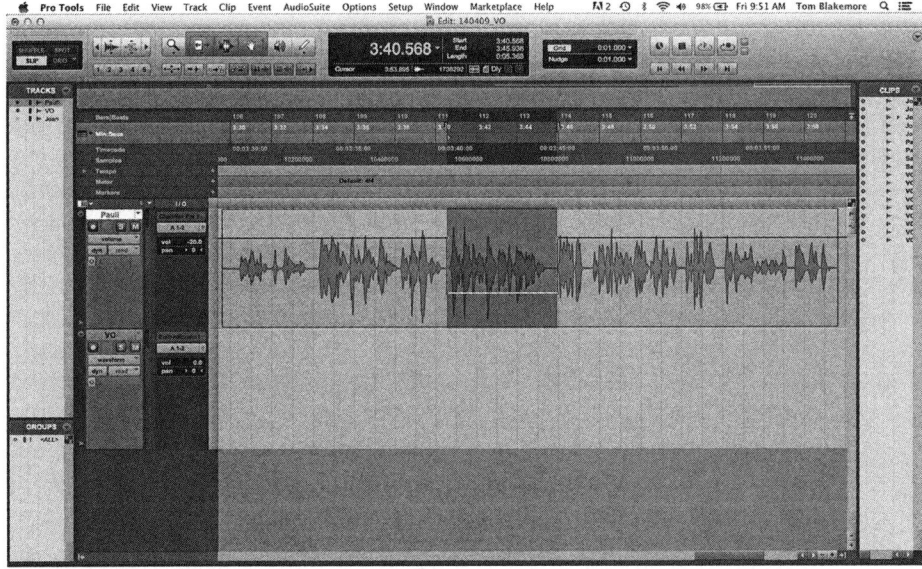

FIGURE 9.5

Pro Tools Volume Graph

circumstance, consider this technique for a good result: First, play the original read for the actor, so that she can get a feel for overall tempo and voice. You may have to play the original over a few times to get it into the narrator's head. Then, while listening to the original read, have the narrator read the new line out loud, to make sure the new words fit into the allotted time (this is all before recording, of course). One good way to check for proper timing is to count the number of syllables in the original and in the revised copy, to ensure that they are reasonably similar. If you are recording into a DAW, you can recall your original processing plug-ins and volume automation. Because you did a good job of documenting the original session, you know what microphone you used, and you have that ready to go. Alright, everything is sounding pretty good, and now it's time to record some takes of the new line. Do a "Save as" command, use today's date to create a new session that is a copy of the original, and open a new track to record on. Then, using your volume automation, lower the line that is to be redone by about 20 dB or so; this will probably vary depending on the talent's preference for how much of the original they want to hear in order to keep the timing the same.

You can now back up a couple of lines in the copy and begin to play across the original reading for the talent. Have him read along to pick up the tempo and feel of the original, and, when the new line comes up, punch into record. Once the line has been recorded to everyone's satisfaction, you can completely mute the original and perform the mix of the new version. This technique is also used if there are to

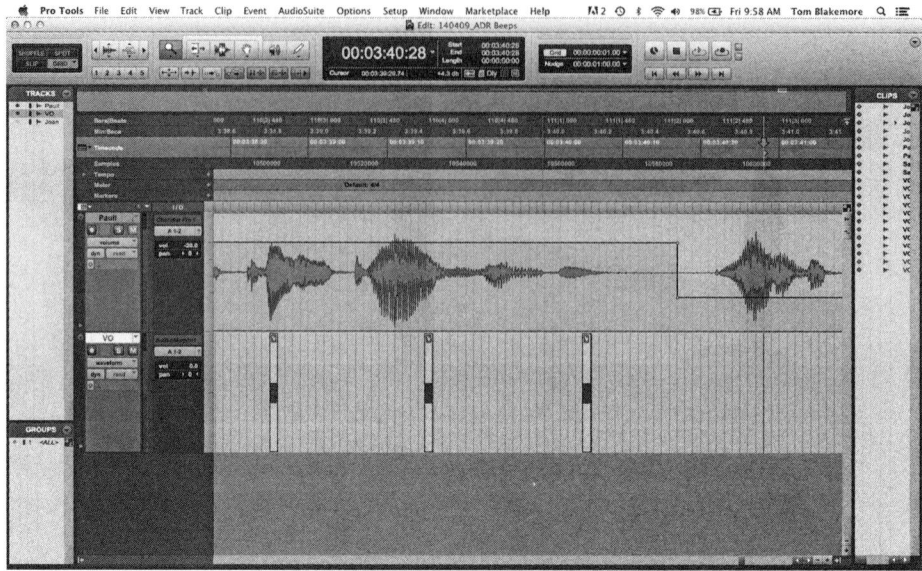

FIGURE 9.6

VO Session With ADR Cue Beeps

be multiple lines recorded to a base spot; as mentioned above, record the longest variation first and use that as the reference, adjusting the speed of the talent's delivery to match the hole into which the new line will go.

What happens if the line that is to be replaced comes at the very beginning of the spot? If there is a music bed to the original spot, you can play that for the narrator and let her know that she should begin reading on the third beat of music. Alternatively, once again we can use an ADR technique. ADR is very often recorded using a series of beeps to cue the actor as to when to begin delivering the line. Three beeps are heard, and, where the fourth would be, the actor begins speaking.

Once again, lower the original line by around 20 dB and place the three beeps so that, where the fourth one would fall, the first line of copy is begun. It's easy to make a set of ADR beeps; use a 1-kHz tone cut to one frame for each beep, and place four of them at twenty-frame intervals. Place the fourth beep directly over the beginning of the first word of copy and then mute or delete that beep. Of course, you could also use this system at any time during the spot for cuing the new line of copy, and some narrators may prefer that, rather than just hearing the original read to cue by. You could make a set of ADR beeps and keep them in a folder for use whenever you may need them, thus saving the time of having to make them each time the need arises. On my DAW, I have a folder that I created and named "Toolbox." In it are a :30-commercial 1-kHz setup tone, a set of ADR beeps, room tone of the studio recorded with my most commonly used mics, and the like.

EDITING AND INTERCUTTING

It's often been said that a movie is made in the editing suite. I think that the same is true of commercials and other audio works. Fitting things to the proper time, selecting the bits that will make for a smooth and effective finished product, and knowing what can be left out are all important skills to develop. Why would we record forty or more takes of a :30 commercial? Are we just hoping to get that "golden" read, all in one magical take? Well, of course, everyone would like to nail the performance in one take that couldn't possibly be improved upon, but that is rarely the case in the real world. Just as with a motion picture, the final product is most often a combination of many different takes, cut together into one complete, final edit and giving the illusion of a flawless and powerful performance. And, as with motion picture dialogue editing, we are often editing at the syllable level (or less!) to make this illusion believable. A skilled producer knows what he or she is listening for and marks the script accordingly, and a skilled engineer is able to stitch all these disparate pieces into a coherent whole, to arrive at the producer's vision, ending with the "golden" take that is the goal. Both the producer and the engineer must have a very clear idea of the process and what they are listening for to make this technique work. Perhaps in no other form of voice recording do we do this more than we do for commercials.

I've heard that, in the US, film editors are called "cutters," and in the UK they are known as "joiners." I think that this is a significant philosophical distinction and I urge you to think of yourself as a "joiner" in your audio work, that is, putting various pieces together to reach the end product, instead of cutting out what you don't want. By keeping very close track of what you are recording, by keeping your ears open to what will go with another piece or not, and by constantly reminding yourself of the time limitations of the spot, this process becomes a creative task on which you can pride yourself.

Figure 9.7 is a copy of a script from a radio commercial with notations for which takes to put together to create the whole read (again, refer to Figure 7.2 for the take sheet from this session). You may find that a particular edit may not work exactly where you have it marked; try moving the edit point a word or a syllable backward or forward and try the edit there, to see if that might smooth things out.

During the recording of the spot, many times either the talent or the producer might call for reading a particular line a number of times, with slightly different emphasis on each. The most common way to do this is known as "three in a row." On each individual take number (slate number), the talent will read the line three times, and the take sheet will be notated as "A, B, C." You can see on our sample script in Figure 9.7 that the last line of the script was recorded a number of times (slates 12–14), and each read was performed three times (A, B, C). The preferred read comes from Take 13B. Intercutting this line into the final product is fairly simple: Just make sure that levels and energy match with what precedes this insert, and the edit becomes invisible to the listener. This, of course, is the goal of all

WRENCHMASTER

"Summer Sale: :30 Radio

H19L5430R

MP3 to Sharon-ASAP

" Mark @ wrenchmaster

SUMMER MUSIC UNDER

NARR (:25):

(7) It's that time again for summer vacation with the family, *and* trips to

Grandma's house, and ~~afternoons at the water park~~ *days* ~~300~~ *(10)* But is your car ready for

summer? Come in now for Wrenchmaster's Summer Sale and get a five-

point brake inspection for free, plus 20% off pads and rotors. And don't

forget a summer oil change – with a new filter for only $22.95.

Make sure that your summer travels are safe and problem free – only at

(13 B)

Wrenchmaster! ~~It's summ~~ertime – *relax*!

TAG (:05):

(8) Sale prices good June 15th through July 20th. Not good with other

offers or sales. Not valid in Arizona or Louisiana.

FIGURE 9.7

Scan of Commercial Script With Takes Noted for Various Pieces

editing—the invisible hand of the engineer creating the "perfect" read from a series of individual performances.

A special case should be mentioned when discussing editing the commercial: The comedy spot. As any comedian will tell you, comedy depends on timing, and performers work for years to develop this particular skill. As editors, we must be aware of the timing of the delivery and how it affects the comedic performance of the spot. Comedy spots almost always run long initially for this reason, and, as the editor, you should be aware of this. Not only is the final performance in your hands, so is the total length of the finished edit, and, with commercials, we cannot end up too long. I'm reminded of the old Jack Benny routine: Jack was known as a notorious tightwad when it came to his money. He's confronted by an armed man who demands, "Your money or your life!" Jack pauses . . . and pauses . . . and pauses some more, without a response. The robber again demands, "I said, your money or your life!" Jack comes back with, "I'm thinking—I'm thinking." The longer the pause goes on, the funnier the bit becomes. If this routine was in a commercial, the pause alone could take a third of the total time. As the editor, what would you do in this situation? Cut the pause down too much, and the comedy isn't as strong; let it play out, and your :30 spot now becomes 38 seconds long. This is one of the reasons that writing for comedy spots is so challenging, and why many of these types of commercial don't work so well. As the editor, you *must* be aware of the goal of the performer and the goal of the advertiser at the same time—a very challenging task for you to master.

There was an instance that I ran into a number of years ago that put all of my skills to the test, and it was an exercise in comedic editing. The advertising agency for the *Chicago Tribune* hired the famed Second City comedy troop to improvise on the subject of the various sections of the newspaper and why they read each of those sections. Three or four cast members would show up for the session, the producer would announce, "Take off on sports sections in newspapers," and I would roll tape. The cast members then improvised their hearts out for 40 or 50 minutes at a time, finished their coffee, and left the studio. The producer and I then had the interesting job of pulling a number of 30-second commercials out of everything that had been performed. Gathering this line and that, we would put together a number of test edits to see what worked and what didn't, and then, if we liked the direction one of the edits was going, try to figure out how to cut it down to the desired time frame. Sometimes, the test edit started out being 60 seconds or longer. What to keep and what to lose? How to keep the comedic timing so essential to the original delivery? Can I cut out a few words and still have meaning and clarity in the sentence? Do I have room to lengthen a pause slightly for effect? Entertaining? Yes. A nice relaxed editing session? Not so much. This particular campaign lasted a full year, and we recorded new material once every two weeks or so; after a time, we began to find our rhythm, and the editing went more smoothly, but the editing sessions were always full of drama and long discussions about what was funny and why, as well as what individual lines suited the newspaper's message. Not only did I have to pay close attention to the overall length of the final spots, but also to what lines

I could intercut with others to make a smooth finished product. Because the actors were improvising, there were no take sheets to reference, only a rough approximation of the time in the recording a favored line might be found. But I have to say, I've rarely laughed so hard during recording sessions!

AFTER THE SESSION

So, now the commercial has been recorded, the sound effects and music have been mixed in, and the producers and clients are all happy. Time to move on to the next session? Not so fast: There is still plenty to do to make this job complete. I've mentioned many times the importance of proper documentation, and it's particularly vital with producing commercials. You never can tell when something is going to be redone or something about the original spot will change. There are many legalities that have to be met so that everyone involved gets properly paid (including you). Any music that you've used must be licensed with the proper entities and reported correctly concerning the total number of spots in a package, length, where they will air, and the proper code number of that piece of music. The actors must have a report filed with the union on session length and the numbers and titles of the spots recorded. The finished commercials must somehow be distributed to the broadcast outlets that will air them, and all of the internal paperwork for use by the studio (or for your records) must be completed and filed.

Commercials are assigned code numbers by the producers, and all communication with the broadcast outlets about the individual commercial uses this code number. Your finished spot's audio file should be named with the code in its title. A mistake in this naming convention can not only cause great confusion in the advertising agency and the station, but also possibly result in the spot not running in the time that has been purchased by the agency. Each spot is assigned an individual, unique number, and it can be thought of as the serial number for that individual commercial. It can't be emphasized strongly enough that any records of the spot and how it is archived contain this code number. If the agency must communicate with you to pull up an old commercial to revise, or to make further copies of it, in most instances they will refer to the code number so that you pull the correct spot. Also, in reporting music usage, you are required to provide this code, and the actor's union keeps track of where and when the commercial was aired (for residual payments) by the same number. Pay strict attention to keeping track of this code number for each commercial that you are involved with. This number is the birth certificate of the spot and will follow it for the rest of its life. Copies of all paperwork should be stored and available at the studio or in your personal files.

When it comes to providing copies of the spot, there are many formats that might be requested. These various copies are provided to the client for approval, to the station for use on the air, and possibly to the voiceover actor for his demo reel. Become familiar with common audio file formats and be prepared to provide any format that might be requested. It is quite possible that you will be asked to make copies in varying formats, depending on to whom you are delivering the copy. We might email an MP3 file to the agency for clearance by the legal department, provide a full-bandwidth .wav file to the station for air, and send a CD to the talent for their use. By the way, it should be pointed out that, even though you recorded and mixed this spot, it is not yours to make copies of for individual use (such as on a talent's demo reel); it is the property of the client, and you *must* have their approval—preferably in writing—before you can distribute it.

It is much preferred that the copies provided for air not be compressed in any way, although more and more stations are regularly broadcasting MP3 files of commercials. If you can avoid data-compressed audio, by all means do so. Full-bandwidth files are usually uploaded to an FTP site for the stations to access and then place in their rotation. If it is a television spot, know what frame rate the station is requesting, and be aware of the proper specifications for the layout of the bars and tones and visual slate and countdown at the head of the file. For radio, it is highly recommended that you provide a voice slate before the spot begins, stating the title, code number, length, and producer (agency) of the spot. For instance, record yourself saying, "30-second radio, Spot Number L-149–67B-18, First National Bank, Acme Advertising," and place this at the head of your file. This way, the station knows that it has the correct spot to air: The agency has already purchased a 30-second block of time and reserved it for L-149–67B-18.

ONE FINAL WORD

Recording commercials is an interesting exercise in diplomacy on your part. There are often a large number of people in the studio during a commercial session, and you might begin asking yourself, who are all of these people? Why are they all here? The producer, the writer, and account executive, the client, someone's receptionist, the lawyer—how do we sort them all out? They each have an opinion about what they're hearing and in which direction the next read should go. Some may like your choice of a sound effect, and some may not. What to do? The trick in this situation is to figure out who is in charge (I would select the producer above all others)—this can often prove to be quite a challenge, but it is essential that the direction in the session has one, and only one, voice. This is your session, and it is your job to make sure that it runs as smoothly as possible. This can be a big challenge for you, both emotionally and politically, but it is absolutely vital that this single voice comes through. Make sure that all direction to the talent runs through only one person; if the talent is receiving conflicting directions on how to change the delivery, it creates so much confusion that she will never know what is wanted. Also, you can direct any comments or questions that you may have to this one person for clarification. Make sure that this session has a leader! I was in a session recently where there were seven people, besides me, in the control room. They soon discovered that not only was there a talkback button on the console, but there were two remote talkback buttons on the producer's desk. Very shortly into the session we arrived at the situation where someone would press the button and say, "Hey Pete, try another read like this"; someone else would immediately press another button and say, "No, no, not like that, try it like this," and so on—each one convinced that their idea was the best and talking over everyone else in the room. The session was quickly descending into chaos! Before I could say anything, Pete, the voiceover, calmly announced, "Hey guys, I'm going to step out for a cup of coffee. While I'm gone, hold an election and vote on who I'm going to talk to."

Remember our previous Kipling quote? *If you can keep your head when all about you are losing theirs . . .?*

There's an old saying in the advertising business: "There's a reason you never see a statue of a committee in the city park." Keep the session running along smoothly and on time; this is your job, your challenge, and your talent.

Recording Long-Form Narration

10

Although much of the general public thinks of voiceover recording as having to do with radio and television commercials, there is a large amount of production being done that requires some particular skills and that can provide a great deal of work for those involved. This is the world of long-form narration. Long form can be considered narrative in nature and is not aimed at selling a product or service (at least, not directly, for the most part). The long-form narration is quite often informational in its intent and aimed at a more select audience than the broadcast commercial. Long-form narration can be classified as corporate (also known as "industrial narration") or documentary. Any type of voice recording that is longer than about 60 seconds can be termed long form.

Corporate (industrial) jobs can include:

- training videos (a very large category);
- material to be used by sales staff to educate potential consumers about a product;
- sales meeting announcements and programming;
- point-of-purchase displays;
- corporate messages (such as a message from the company president).

Documentaries can include:

- film and television narrations;
- radio narrations (such as longer news stories and more);
- use on the Internet that does not fall under the heading of "Corporate."

You should be aware that this type of recording is still telling a story—whatever that story may be. We should always be mindful of the pacing and flow of the narration, so that our audience remains interested in what we are telling them, however dry and boring that information may be. Imagine the skill it takes to read a 15-minute piece on a new dermatological drug and sound interested and excited about this new product, all the while effortlessly tossing off Latin medical terminology as if you spoke this way every day! Telling the story to hold your audience and having them stay interested and excited about what you have to say is what this type of voiceover is all about.

Everything that's been brought up in this book till this point still applies: Working with the talent and client, your microphone choices, and so on, are no less important when doing long-form sessions or any other type of recording. Developing a relationship with the people you will be working with is very important and goes a long way to developing the trust that is paramount in having a long and satisfying career. As I've mentioned previously, it is from sessions like these that I've made and kept lifelong friends, and these relationships often go beyond life in the studio and become part of the social fabric of our lives. But it all starts with your professionalism and aptitude for handling your job effortlessly and with a sense of humor.

Some years ago, I worked with a director on many occasions, and the director always used the same voice talent for all of his jobs. The director was originally from Germany, and both he and the actor were getting along in years. It was obvious that the two were very close personal friends, besides having a working relationship, and the warmth between the two men showed through in all they did. After working with the two of them for a year or so, I happened one day to ask how they met and formed such a close bond. It turned out that both were World War II veterans, and, when they discovered this, they began discussing where they had served during the war. It turned out that not only had both fought in the Battle of the Bulge (on opposing sides), but that they had both been trying to defend and take a particular village—they had been in the same battle, on the same day, and had been shooting at each other! On learning this, they realized the folly of war and became lifelong friends. So, out of narration sessions, relationships are made and kept.

Long form is usually not done by advertising agencies (who do the commercials). There are companies and producers who specialize in this type of work and are very familiar with the intricacies of the long form. Many times, the production is done by the company whose product or service is being talked about itself, either using in-house writers and producers or utilizing freelance talent or one of the specialized companies for these services. There is a difference in approach that must be taken into account when writing for a :30 spot versus a long training video: The approach to language and how we address the audience changes. We have a longer time frame in which to get our idea or our message across, and we're not selling anything (per se). Once again, the content is more informational.

You will find that the session pace is more relaxed, not because what you're recording is any less important, but because the budgets and time considerations, both for the finished production and for a drop-dead deadline (time to air), are not as stringent and filled with anxiety. The work has a more relaxed flow to it.

It's been my experience that a company doing one type of work (advertising or corporate) usually has little understanding of the other world, because both are so specialized. There are, of course, exceptions, and, in smaller markets, companies have to do both to survive. However, in larger market areas, you will usually find an either/or situation as to what type of work a company does. Also, generally speaking, the corporate-communication firms tend to be smaller businesses with fewer employees.

One of the benefits of doing corporate long form that I've enjoyed over the years is learning some interesting (if not entirely practical) information: Through these sessions, I've learned how spark plugs are manufactured, how to properly insulate a house, the history of one of America's largest beer breweries, how to correctly cook the French fries in a fast-food establishment, why earthworms don't like the light, how Afghan rugs are hand woven, and so much more. If you want to know about the migratory habits of the American bison before the white settlers, or what Thomas Jefferson has to do with the way our education system is structured, I'm your guy. As you might imagine, I'm always ready to talk about the most obscure subject with anyone at a party! Like I said—interesting, but not particularly useful in day-to-day life. Folks involved in long-form work are veritable fonts of useless information.

Many people not involved with the corporate, or industrial, type of work look down on it because it isn't sexy and exciting. Consider this, however: We once were tasked with producing a video piece that played at a sales conference for an auto-parts supplier. The video was to announce the prize for the top sales person in each region the supplier did business in, and the grand prize was—two weeks in Monte Carlo! Well then—the producer put together a film crew and they spent ten days in Monaco and the Riviera, tooling around in a brand-new Ferrari and sampling all of the great food and hotels available in order to illustrate the lifestyle of this glamorous location. Hmmm—not an exciting way to make a living? I'll let you be the judge of that. By the way, the video was a success, and the client was so pleased it came back to that producer for years to put together all of its trade-show, sales-conference, and point-of-purchase materials. Although this may be an unusual job, opportunities do exist to get out of the studio and travel when working in this particular industry. We have to shoot footage wherever our client might be doing business, so it's time to pack up the location gear and hit the road.

Unlike the commercial session, in long-form narration, we have more latitude in changing the wording of a sentence to make the meaning or flow clearer. Yes, there will be legal issues that must be taken into account, but, for the most part, we can change word order or substitute one phrase for another, if it helps the audience better grasp the message. Because of this, engineers and voice talent involved in the long form learn to be very good editors (even if they are not writers); having someone who is hearing the wording for the first time in the session is very useful to the writer of the script, and these people can usually point out unclear or cluttered sentences. Just as the audience will be hearing this for the first time, so are those in the session who are not familiar with the script beforehand. This is one of your main jobs in a long-form session, and you should know when it is permissible to speak up and suggest a correction to the wording.

As has been pointed out earlier, diplomacy is an important skill to develop to survive life in the studio, and this is but one example of the importance of it. Knowing when to speak up and who is accepting of advice is an intuitive thing to learn, but remember that the writer or producer is there for the same reason that you are, and that is to arrive at the best possible production, and so they are usually open to your suggestions. I've found that the producers of long-form narrations usually don't have quite the egos that commercial producers have. I think that this may be attributable to their not being under as much pressure from time and budget as their advertising-agency counterparts, and so the sessions tend to be more relaxed in feel. Be very aware, however, that the finished product is still important, and that no one involved takes this lightly.

One thing that's nice about the long-form narration is that, in most instances, there aren't the time worries that come with commercial recording. Whether a piece runs 32 or 41 seconds is usually not an issue. It takes as long as it takes to come to a clear understanding of the content of the script, and, if a longer pause or a few additional words are inserted to help get there, most times it is of no consequence. The exception to this would be if the picture has already been edited and the new narration must fit into the time frame of the picture edit. If this is the case, the script will probably be marked with the timing of the paragraphs or sentences, so have your stopwatch handy and make sure that the narrator doesn't run long. However, once again, if the read is too long, there is the latitude to choose different wording to help bring it in on time, while still maintaining the overall meaning.

Because the scripts are longer than for a commercial, the session length is longer, as well. You must pay particular attention to the possibility of talent fatigue, and, if you sense that the narrator is losing focus, or that the voice is becoming a bit frayed, call for a break to give them a chance to recover. Some long-form sessions can run half a day or more, so this is a real concern. Also, know your own limits if you are engineering; you are being counted on to stay on top of things, and, if you find your own concentration wavering or if you find yourself starting to turn up the monitor speakers, by all means request a break. Generally, everyone will welcome a short time to stand up, stretch, make a phone call, grab a cup of coffee, and so on, and the result clearly shows when it's time to resume recording. It should be pointed out that ear fatigue is real, for both the engineer and the talent monitoring on earphones: Playback levels tend to creep up and up, and the frequency response of your ears begins to lessen. Know what this symptom is and be mindful of it in the studio. Besides being an issue for your long-term hearing health, the final product can be degraded if those involved push themselves too hard. Breaks can be an incredible benefit to the overall outcome of the job; just be sure to keep your concentration sharp during and after the break and keep the energy level up in the studio.

THE LONG-FORM SESSION

So now then, what is our approach to recording? We still want to address the audience one to one, unless we are providing announcements at a sales conference, for instance. Microphone choice is still important, as is consideration of the final playback situation. This is something that you should always be aware of—it can vary greatly, and you should assume nothing. Your choice of mic should not be compromised for doing this type of work: Give it the same amount of thought that you would any other type of recording situation. Having two actors in the studio at once is not uncommon, and they will trade paragraphs, or perhaps one voice is a character voice of some sort. A pairing of a male and a female voice is common for the variety it provides. For both the engineer and the voice actor, long form provides a chance to stretch out a bit more than with commercial production. You are not constrained by a strict time limit, and the longer script gives you more room to play.

As has been pointed out previously, the script should be prepared with phonetic spellings of unfamiliar words and proper names, which is especially important when doing a medical-type reading. Some medications and procedures are virtually impossible to pronounce unless you have been to medical school, and breaking the word down into simple syllables is a great benefit, not only to the narrator, but to the engineer as well. Imagine reading a script cold (that is, without having practiced it)—not an unusual thing to happen for a corporate or documentary session—and running across "Hydrochlorothiazide"! The poor announcer is making his salary today! However, if you substitute "Hi-dro-clor-o-*thigh*-a-zide," the word becomes much easier on the eye and is easier to pronounce. This is one of the reasons it's also important to have the script double-spaced—it allows room on the page for the talent to make notes on pronunciation and emphasis. Even with the phonetic spelling, it will probably take a few readings for the narrator to roll across this word effortlessly, but not nearly as many as it would without the phonetic guide. Remember, your audience is going to be familiar with the word, and so the narrator must sound as if he is also, just as if he uses this word all the time in normal conversation.

The same is true for people's names: You should be able to say names as if they are your best friends. For instance, most people pronounce my last name as "Blackmore" when they first encounter it, as this is a more common last name, but the proper pronunciation is "Blake-more," and you wouldn't want to end up announcing the president of a company and mispronouncing his or her name. So, please, give a phonetic guide to help everyone along. The goal of the long-form narration is to sound as if you are the authority on the subject at hand, and proper pronunciation and not stumbling over words are a big part of that.

Double-spacing of the script has another benefit for the narrator: Most announcers "read ahead," that is, their eyes are a sentence or so ahead of the words coming out of their mouth. By doing this, the announcer can anticipate where the sentence is heading and make better sense of it. If the script is double-spaced and doesn't have everything in capital letters, it becomes much easier on the eye and makes it easier for the narrator to navigate his way through the script. Reading ahead allows the narrator to anticipate upcoming problem areas, or perhaps spot a word that may need additional emphasis. You'll be doing the announcer a big favor if the script is formatted properly in this manner. And, while we're on the subject of formatting scripts, it is of the utmost importance that, if at all possible, you don't have a sentence beginning on one page and continuing on to the next. This makes it very difficult for the announcer to read ahead and is also an invitation for noise to creep into the session in the form of paper-shuffling sounds. If the narrator has to turn the page, she can pause between sentences, turn the page (allowing you a clean edit point to get rid of the noise), and then continue reading.

Most experienced voice actors have developed a skill for turning the pages of the script silently as they read. It often involves lifting off a page, reading it, and then, when they are almost at the bottom of that page, lifting the next page and either dropping the first to the floor or placing it silently on the copy stand. This technique has a couple of benefits for us: The first and most obvious is the lack of paper noise being made, and the second is that it allows the actor to hold the script at eye height or just slightly below while they are reading. This keeps the head fairly straight and forward facing, and the narrator isn't reading down (and, thus, away from the microphone). If the script is left on the stand, by the time the actor is at the bottom of a page, the angle of the head becomes such that the mouth is nearly 90° off mic. As the copy is read, the head is moving up and down, and the resulting mic position becomes very apparent. When it comes to silent page turning, you, as the engineer, should become proficient at it as well, as mentioned previously: If you are making a lot of paper noise in the control room, you will mask the noise coming from the studio and you may well miss marking a spot where you have to make an edit to clean up the paper shuffling by the actor.

Another strategy often seen has the narrator folding the top of the script over the copy stand, so that it is more in line with the eyes and not sitting on the shelf of the stand. This does alleviate some of the head movement and is a much more comfortable angle for the actor.

As discussed in Chapter 7, The Session, there is a question concerning the talent's comfort that revolves around whether he should be seated or standing. Owing to the longer length of this type of recording, a stool should be provided, so that the talent has the chance to get off her feet at least for a little while during the session. The longer the script, the more obvious the need to give her the choice of remaining seated for the entire read. However, be careful that the airflow from the diaphragm and up through the mouth isn't constricted. For this reason, I'm not a fan of having the actor seated at a table, although many narrators prefer this. In the end, I'll leave the choice to them, but, if they choose the table, I'll listen very closely for a pinched or choked delivery and remind them to sit up very straight (once again, we're facing the question of your diplomacy—you don't want to sound like their mother, always telling them not to slouch and to sit up straight!).

DOCUMENTATION

Once again, as with all types of recording, we come to documentation and the keeping of effective notes. As with commercial recording (or any type of recording, for that matter), you should note the type of mic and any processing that you have engaged for the recording. Because of the length of the script and the overall length of the finished product, using a take sheet just isn't very effective. Making notes directly on the script provides the clearest and easiest way to keep track of the takes and any other notes that you need to make. Refer to Chapter 7, The Session, for an example of a long-form script and its associated marked copy. If you are doing pickups or insert lines, it is much easier to mark them directly on the script itself and much more effective when it comes to the editing portion of the session. Proper documentation is critical in the event that you will have to do any voiceover recording after the initial session, which could be the next day or even years in the future. For instance, back in the days when cable television was just starting to make inroads into American homes, each community would grant a legal monopoly to one or two cable companies to operate in their town. As part of the licensing procedure, a number of cable companies would appear before the city council and give a presentation on why their particular service and pricing would be the best for the community. I did a lot of work for a company whose client was one of the national cable providers, and we put together an audio-visual presentation to show to the city fathers. This presentation was structured in such a way that roughly three-quarters of it never changed—we recorded it once, and that was that. However, in the middle section was a portion customized to each locality: city name, a few interesting facts about the town, pricing options based on population, and so on. The beauty of this presentation was that each city thought that we had put it together just for them— a lot of work to go to just for a preliminary pitch to the council. In order for our illusion to work, it couldn't sound as if the narration had been done in two different months in two totally dissimilar rooms. This required very careful matching of the narrator's reading and recording, and it was the original documentation that was the key to pulling this off. This particular project went on for over four years and was very successful. Imagine recording someone three years after the original session and having your match make or break the presentation. Such is life in the long-form world.

Part of your documentation should, of course, be a listing of all music and sound effects used and the proper filing of licensing materials with the copyright holders. Quite a bit of long-form work is done as a one-time use (unlike a commercial, for instance), and there is the temptation to avoid paying licensing fees. After all, the presentation will show once, and no one will ever know what music you used, right? You would be surprised how often representatives of the American Society of Composers, Authors and Publishers or Broadcast Music, Inc. attend events looking for just such violations of the copyright code. You only have to get caught once to put your company out of business, such are the fines that the courts can impose. So, be on the safe side and do your licensing, please. One last word on this subject: These licensing fees pay the royalties to the composers and artists who provide your music, and it's how they make their living. You expect to be paid for your work, and these folks do as well. Any company that provides music for stock usage will easily provide you with its payment schedules, so there should be no surprises if you are honest about what you want to use its music for.

As with commercials, there is a standard union rate structure for the voice talent through SAG/AFTRA. You, or a studio representative, are responsible for keeping accurate records of session length and date(s), in order for the actor to be paid fairly. The good news on this front is that the hourly minimum charge is less for long-form work (again, termed corporate or industrial work) than it is for commercials, even though the session length is generally expected to be longer.

Needless to say, proper documentation of all session elements and their proper storage is key to having this process work smoothly and efficiently. You should have or develop a system and stick with it for all future filing and documentation. It was through this documentation process that we kept a project on track and running smoothly for a period of years. We had put together a film that was shown to tour groups at the beginning of the tour of Miller Brewing in Milwaukee, Wisconsin. Over time, its product line changed, or it came out with a seasonal brew, or perhaps simply wanted to update the film. We revised and revised so many times that I lost track of the number of versions, but the new sections always cut into the original seamlessly. In fact, the biggest problem we had was that the narrator was getting older, and his voice started to change! The Miller folks finally pulled the film from their tour, partly because the narrator was now starting to sound like an old man.

RECORDING TO PICTURE

In any type of voiceover recording situation, you may be called upon to have a video reference to record to. Usually done for timing purposes for an already edited picture, recording to picture can introduce its own set of challenges, but also can be a terrific tool in the session process. Whether commercial or long form, recording to picture ensures that your timings are correct; it takes the stopwatch out of your hand and allows you to more fully concentrate on the performance. However, there are a couple of things that we should be aware of. To begin with, recording to a visual reference is something that the voice actor must practice to be skilled at. The first couple of times a narrator attempts to keep one eye on the script and the other on a video monitor, you can expect some missteps. However, as time goes on, the process becomes more natural, and the sessions begin to flow more smoothly.

In setting up for a session involving picture playback, make sure that your visual playback chain is as clean as your audio signal path. It becomes quite distracting to have an image that flickers, becomes pixelated, or drops out when you are trying to record, but, believe it or not, it doesn't really matter if the picture is black-and-white or color or if the resolution isn't full HD quality. Those things aren't the point—the timing is what it's all about. Walter Murch tells the story of mixing *Apocalypse Now* to a reference video on ¾-inch tape and in black-and-white. The first time he saw the film in color was at the premier! In the days of film, processing was costly, and color processing even more so; to provide a black-and-white work tape to the audio department made all the sense in the world and didn't slow anyone down. Today, in the age of computer production and Quicktime movies, this process becomes faster and more cost-effective, and there simply isn't any reason not to have a quality picture to work from.

You should *always* give the picture editor (or whoever is providing the visual materials to you) a list of specifications that you'll need to do your job. This is part of the communication and documentation process essential to an effective workflow. Your specs should include file type—*and size*. Too large a video file can slow down and even crash audio workstations, and so we most times request something less than full resolution if it's a long project. Of course, audio layout should be discussed; for instance, is the picture department going to be providing you with a separate OMF or AAF file with existing audio, or will a stereo track be sufficient for your needs? If they are providing more than two tracks of audio, do you have a preference as to layout? How about track names? Give these issues some thought when requesting what you are going to need for a smooth, trouble-free session. Then, when you receive the materials, take a moment and check them against your specs—send them back to be redone if they are not correct (don't hesitate to send them back if need be; after all, others will do the same with your materials if you get it wrong). There's a bit of a controversy these days on whether or not you should include a time-code burn window on the picture. A time-code window shows running time-code, usually located in the bottom quarter of the image, to use as a visual reference for sync and timing.

One school of thought has it that we just don't need it anymore, because we have a continual code readout on our digital workstation's main record window. True, I suppose, but I always request a burn-in for multiple reasons. The first is that you can easily check sync by matching the time-code number you see on the screen with

FIGURE 10.1

Time-Code Window Burn

Source: Courtesy of Tribeca Flashpoint Media AA; photo by the author

the number on your workstation's display. The second is that a visual indication of the time, flowing across the screen, gives an actor (or Foley artist) a cue point for when to begin delivering a line or performing an effect. It's a "count down," if you will. It also allows the actor a reference on how long they have to complete the line. I've worked with narrators so accomplished at recording to picture that, once everything was lined up in my workstation, we would drop into record, the actor would look to see at what time-code number his first line was to be delivered, and away we would go, working our way through the script, with his lines delivered exactly where they were supposed to be. If a mistake was made, or if the timing needed to be adjusted, we could back up, roll into that point, punch into record, and redo the line. At the end of the script, all of the lines were in place, and no editing needed to be done! Talk about time-efficient. Of course, the script had to be prepared in advance, with all of the time-code numbers clearly noted for the actor to begin in the right place, but this time spent was more than made up for by the time saved in the session itself, recording and editing. My advice to you is always, always, always request a time-code window with your picture and have the picture file remade if it's not done correctly.

On the subject of delivery specs, just a quick word about one other thing that you should always request be included, and that is a sync pop. Also known as a two pop, sync bloop, and other names, this is an aural cue as to proper synchronization between picture and audio. Placed *exactly* 2 seconds before the first frame of picture and lasting for *exactly* one frame, the pop lines up with the 2-second mark in a visual countdown on the picture. By visually lining up the waveform of the pop with the "2" frame of the countdown, we know that both picture and audio elements are properly aligned. Coming from the old film tradition, the sync pop is still a useful tool today. With long-form projects, you might also request an end pop, 2 seconds after last frame of the picture, but this is usually done only in the film world and not necessarily for commercial or long-form narrations. However, it's a nice reference to have and gives you peace of mind that your sync hasn't drifted during the course of the program material. Utilizing the end pop, if a question of sync comes up later, you can always prove that everything was in sync when it left your hands.

As mentioned in an earlier chapter, you have to make sure that your actor can see the video monitor clearly; you don't want it blocked by the microphone or the copy stand. This may take some repositioning of the microphone on your part, or possibly moving the actor a bit in the studio. Remember that our number one goal is to capture a clean signal, and the microphone positioning can't be compromised, but please give the narrator a fighting chance of seeing the monitor. Also keep in mind that, besides looking at the monitor, she will be reading from her script, so that has to be positioned correctly as well. Although getting everything situated in the studio correctly certainly isn't difficult, you have to keep these things in mind. Remember our motto here: The talent's comfort is paramount to getting a good performance.

You should check with the talent as to not only the preferred headphone level that you are sending, but also the balance between his own voice and the track coming from the video. Ask beforehand if the talent prefers a single earphone or the more common pair, open or closed design, or if no headphones at all is the preferred style of working. Again, it all comes back to the talent's comfort and your doing anything in your power to achieve it.

FURTHER THOUGHTS ON LONG-FORM NARRATION

As with any other type of recording, you should always consider your ultimate playback situation when recording long form. Unlike commercials, which usually are played either on the radio or television, long-form narration can be played in an endless number of scenarios. From acoustically designed and perfect theatres, to reverberant hotel ballrooms, to the tiny speakers on a laptop computer, each must be designed into your sound. For instance, to give a little "oomph" to the VO, I sometimes will add just a touch of reverb to the voice. However, if I know that this will be played back in a reverberant venue, I can't afford to do this, because the natural bounce in the room will be added to the reverb that I have added, and I stand the chance of losing clarity in the narration. The same is true if I know beforehand that the piece will be played primarily on laptops or other portable devices: I can't add in a lot of extra low-frequency information, or the voice will become muddy. Finding out the intended use of what you are recording is vital in any recording situation, and the long form is no different. Make getting this information ahead of time one of your priorities. Learning to recognize and avoid these potential problem areas is one of your jobs, as is learning how your particular audio monitors will translate into a number of playback situations. If possible, it's recommended that you have a number of different monitors connected, to check how things are going to sound in the real world.

When it comes time to edit your raw voice tracks, there is the trap of having so much material before you and so many takes that wading through it all becomes a real chore. This can test your powers of concentration. You may have to move a lot of information around because it was recorded out of order (pickups at the end of the session, for instance), and it's easy to make mistakes. Also, owing to the large amount of information that you are dealing with, it's possible that your concentration

FIGURE 10.2

Time-Code Window Burn

Source: Tribeca Flashpoint Media AA; photo by the author

can wander. You certainly don't want a repeated sentence or deleted word to pop up in the final, released version. I have a couple of recommendations that might help you out in these times. If you recall, in the chapter on commercial recording, we talked about letting your workstation work for you with its auto-numbering system to help you find individual takes. The same is true with long-form narration. If you carefully note your takes, and are careful with the auto-numbering function of the workstation, it is relatively easy to find the correct takes and edit them into the proper order. Just be sure to check your edits as you make them, to ensure that you haven't clipped off part of a breath or included a double breath, and that the words are in the proper order. Then, when your editing is completed, listen back to the entire narration while reading the script at the same time. Although this may be time-consuming, the final quality assurance is what separates the casual engineer from the exceptional. Remember, editing is part of the session also, so give it the attention to detail that it deserves.

The world of long form has been my bread and butter for many years and has served me well, both in terms of income and in terms of personal relationships. The chance to work in a relatively low-key atmosphere, while learning about the most esoteric subjects, isn't my idea of work. It's not sexy, but it is a hell of a lot of fun!

Recording for Games and Animation

11

Most of the recording situations that we've discussed so far are quite similar in strategy and technique; whether recording for commercials or long-form projects, the only real difference comes with the approach to documentation and the length of the session. However, when we begin recording for the worlds of game audio and animation, the terrain shifts, and we have to start thinking about new ways to go about our task. There are a couple of major differences that we must consider. I have included both game and animation voice recording in this single chapter, for the simple reason that both of these forms are animated. As well, they share certain key recording-technique features. However, in reality, there is an underlying and fundamental difference in philosophy of the two types of voice recording, and this should be kept in mind as you go through this chapter. Also, recording for games and animation generally takes longer than recording a straight narration: When the talent walks in, the question for a narration might be, "Would you like this read seriously or a little lighter in tone?" For games and animation, considerable time can be spent in trying to find the right character for the voice, and fine-tuning the performance can be time consuming. Plan for extra time in your session for this process. Also, for both types of recording, we most times will want to have multiple approaches to any given line of script, and this, of course, also takes extra time.

GAME VOICE RECORDING

This section is not intended to be an overview of the entire game creation process, which is multifaceted and involves a large number of people and disciplines; there are many good books and other resources on the subject that you can easily find. I might point you to a brief, but highly detailed, discussion by veteran game audio professional Rich Carle, in his contribution to David Lewis Yewdall's *Practical Art of Motion Picture Sound* (4th edition, 2012, by Focal Press, Waltham, MA). Recording dialogue for games and animation is a unique combination of production sound recording and the post-production process. Almost always, the voice is recorded in the studio, unlike motion picture audio. Because the voice is recorded in studio, a dead-quiet room, free of reflections, is a must for this type of work. Refer to a brief discussion of an alternative method later in this chapter, when I discuss the dialogue recording for the animated feature film *Rango*.

Although I may be accused of stating the obvious (and rightfully so), it should always be kept in mind that the underlying principle of the gaming world is "randomness." Games, by their very nature, are based on the concept of nonlinearity and random access; this brings about a new way of thinking about recording the dialogue for this discipline, and it is fundamentally different than recording for linear media (films, commercials, etc.). During game play, we can never predict, as game makers, what choices the player is apt to make, and, as a result, we must prepare for all (or at any rate a great deal of) eventualities. It is the very random nature of the experience that gives game playing its appeal. As a result, the strategy that we must develop to record the lines of dialogue is quite different than any other type of voice recording. Let me set up a very simple scenario: There is a character (Character A) standing beneath a balcony. On the railing of the balcony sits a flowerpot. The wind begins to blow the flowerpot closer to the edge of the railing, and it's about to fall off and onto the head of Character A. Another character (Character B) sees the situation and calls, "Look out!" To record this little scene for an animation (or a film, commercial, or virtually anything else), we would have both of the character actors in the studio at the same time, interacting in real time. However, in the game universe, it is quite possible that the player never chooses to have the character stand under the balcony at all, or chooses to move the character there 10 minutes from now. In the animation, we can record, edit, time, and mix the dialogue exchange, and the animator will draw the action to our timings. In the game world, we must have the "Look out!" available if the need arises, but no one can know with any degree of certainty when, or if, that line of dialogue will ever be needed. So how to record our dialogue?

Within the game, there are times of story or character development that do not include game play. These are known as "cinematics" or "cut scenes," and they play out like a short film. They are linear and they are not random. For these instances, the dialogue can be recorded just as if it were to be included in a film or animated piece. The key word here is "linear." The remainder of the game is controlled in a random-access manner by the player. For dialogue to be included in these other portions of the game, we must, because of the nature of the experience, alter how we would go about capturing this dialogue. Because we cannot predict when (or if) any of the lines we record will be used, each line of the script must be recorded in a stand-alone manner, without any connection to another line of the script. Because of this, we must pay particular attention to our documentation process (remember that from an earlier chapter?) and to the script. We will record only one character at a time, instead of having real-time interaction between the actors. Once we have worked our way through the script and recorded all of the dialogue lines from one character, all of the takes are named according to a predetermined protocol established by the game programmer, and each of these individual lines is then passed on to the game developers for implementation into the final product.

In our example of the flowerpot above, we can imagine that Character B could give a reading of the "Look out!" line in a number of different ways. The emotion could be one of dire warning. It could be that Character B actually wants the flowerpot to hit A on the head and delivers the line in a singsong, taunting manner. Character B could be stumbling toward Character A and trying to catch the flowerpot before it strikes. Each of these eventualities will want to be covered so that the game programmer has a choice of the proper way for the scene to play out, and that requires a large number of takes, each of which must be carefully noted to allow the programmer to find the reading and emotion he or she prefers. You can begin to appreciate the complexity for the engineer when it comes to recording and keeping track of all this material. As I mentioned above, all of this may be stating the obvious, but you should be prepared for this type of situation as you go into the session to begin recording the dialogue.

Generally speaking, when recording dialogue for game use, we would record each of the characters individually (that is, having only one actor in the studio at a time). Because this does not allow for a natural interaction with another actor, we have to pay particular attention to the tone of the reading, as, in the game play, the lines we record are probably going to be in response to lines from another character, whether that be in a conversation or some other vocal interaction. This attention to the tone and consistency of the reading is paramount, and it is as much the responsibility of the engineer as that of the director to make sure that all of the individual line readings will come together into a unified whole during game play.

Also, we should be paying particular attention to the naturalness of the delivery by the actor. When we are recording a narration for a training video, for example, we would want the reading to be well enunciated, clear, and spoken in an almost

INSIGHT: D.B. Cooper, Voice Talent, Casting and Directing

Voice actors are, for the most part, eager to please. They're one of the final conduits for the work of so many—the writer, the director, the producer. They are entrusted with the job of breathing life into characters who populate a virtual world; it's a heady mix of challenge and responsibility and pleasure. And actors want to do a great job, not only to please the folks in the studio at the time of the session, but to be called upon again.

Veteran game VOs know that games need to be recorded fast—there's so much script and so little time. Game scripts can have hundreds, if not thousands, of lines. Time and budgets drive the need to get the work done with hyper-efficiency, and your actors will be looking to help speed the plow.

Because actors are keen to help get the work done quickly, they'll often approach lines of script as elements that need to be rendered whole. Trouble is, a script can sound stilted when lines are read as *lines*. Even an excellent actor risks sounding odd when delivering lines that are spoken clearly all the way through; the only places we are liable to find fluency in all exchanges are movies and plays. For the most part, people tend to speak in phrases or fragmented parts of sentences. We breathe, we think, we change our minds mid-sentence. We start, stop, and correct ourselves. We use vocal crutches, such as "y'know" and "uh." So, if you don't want your game script to sound like a performance, allow mistakes and corrections and pauses—as well as fluency.

Players can tell when lines are being "read," and it sunders the suspension of disbelief. No director wants this. Directors strive for authenticity, reality, honesty. You want compelling dialogue that sounds like a real conversation, even though the different sides were recorded days apart. As you begin to explore the script together, assure your actors there is time:

• to find authentic pauses:

> "Great. Now we look incompetent."

becomes

> "Great. [sigh] Now [inhales through nose] we look incompetent."

• to self-interrupt:

> "Sorry guys, I know you're just coming off a case but I'm afraid we're going to have to make a detour."

becomes

> "Sorry guys, we've got to—look, [exhale] I know you're just coming off a case but I'm afraid we're going to have to make a detour."

- to say "uh ..." now and then, like real people do:

> "Jesus—he looked ... uh ... pretty mad."

Engage in character building with short lines from the script, so you and the actor can find where the character "lives" vocally. Allow time for a newer actor to get the feel for different intensities with the new character voice. Creating the character and stretching it out at the start of the session will reduce the likelihood of needing to stop to rejigger the character for whispers or shouts. This keeps the flow going and also keeps your actor's composure smooth and confidence high.

Keeping your actors on track with their characterizations and accents is totally in your court as director. Your actors will be relying on you to be "the ears" during a session. Experienced actors need shepherding a lot less than neophytes, but it's a source of security for everyone behind the mic. Knowing the director is keeping on top of the intricacies of the performance allows the actor the freedom to act and not self-censor.

If you do hear the actor lose their way with an accent, or give a line a read that sounds totally wrong, you may feel the urge to stop them in their tracks. Interrupting the actor may seem like a time-saver, but what it really does is alarm them and undermine their confidence—which can lose you a strong performance, or it will irritate the actor and make them less likely to be receptive. Just be patient! Wait till the end of the line. There are few large paragraphs in video-game dialogue scripts.

Let the actor know, at the beginning of the session, that it's OK to ask what you mean. You probably use a set of terms that make perfect sense to you, but someone who's never encountered them in a recording environment may be in the dark. If you aren't hearing what you want, you need to take the initiative and explain your direction. And, if the actor is totally stumped about how to give you the line the way you need it, that's when a line read is appropriate. You can offer the line the way you hear it in your mind. If it's presented with a spirit of genuine collaboration, it's more freeing than limiting.

Taking time to help your actors to break out of the "good-dog" mode of getting a job done—to relax and settle into their place as actors—gives them the freedom to step fully into their characters and creates a richer recording experience, and everyone ends up with great audio.

For more information, contact: DBCooperVo.com

formal reading voice. However, game audio more closely approximates our casual, less-formal speech, full of "umm"s, awkward pauses, and other vocal tics that would be out of place in a narration piece. The question now becomes how to keep the actor sounding natural, instead of stilted and "theatrical" in the reading. We could allow the actor to imagine the other half of a conversation and respond accordingly, but this is often an unsatisfactory solution to achieving a natural back and forth between the two characters. A useful thing for the actor is to have a "reader" in the studio with him, reading the other character's lines to help with the reaction and inflection in the lines of dialogue, but this is simply not the same as interacting with another actor.

Because our actors will, for the most part, be speaking in a "character" voice and not their own normal speaking voice, keep in mind that this can put quite a bit of stress on the throat and vocal cords. To have an actor continually speaking in an unnaturally low and gravelly voice, or prolonged periods of very loud delivery of lines, or maybe a high and squeaky voice can fatigue the talent in a very short period of time. I've actually done sessions in which the actor put such a strain on the voice that he could barely speak at all the following day! These people rely on their voice for their living, and you don't want them to destroy their instrument, even for a short amount of time. Remember that one of your jobs as an engineer is to keep everyone comfortable, and this situation is a prime example of that.

It's easy for us, as listeners to a performance, to want to keep forging ahead, but this is one instance where taking frequent breaks can actually lead to a more productive session. Pay attention to your actor; if you know the person and their nonverbal communication (another great reason to take the time before the session for an informal conversation about anything at all), you will easily pick up on when he or she is becoming voice-fatigued and needs to take a minute and allow the vocal cords to calm down. Once again, your documentation of the takes and lines should be such that you can easily find where you left off before the break, and it's vital that, once recording resumes, you should take a minute a play back the last few lines that were recorded before the break, so that he or she can find that character again. This way, all of the individual performances will match, and the character will remain consistent. Also, take into account the mental aspect of the talent's job: Trying to maintain a consistent character while also paying attention to delivering a good performance and keeping track of the larger goal of the game or animation piece can put strain on more than the voice. Besides resting the throat, these breaks can also help in keeping the actor's energy up because of the mental break that they give. It's easy to miss the big picture in all of the little details of the session, but, in game and animation work, the consistency of the character is paramount.

In the world of game audio, resources (CPU, bandwidth, disk space, RAM) are limited, and the content providers must always keep in mind that the game has demands other than the audio; art, level design, cinematic videos, programming instructions, and more, are competing for finite space. And yet, as audio providers, we must be prepared to cover all eventualities that the player may encounter during game play. To make the gaming experience as rich as possible, we have to avoid redundant sound cues as much as we can, and, to this end, the dialogue recording process can become quite complex. Often, the same line of dialogue is given to multiple characters to cover the possibility of a commonly occurring action, so that each time this event happens, the player hears a line of warning (for instance) coming from a different character than the last time the line was spoken. Also, there may be times when we want to use the same dialogue line but process it somewhat differently, to give a feeling of distance or of the ambient characteristic of the space the player finds himself in. Often, these changes in environment can be accomplished by the game engine introducing a random effect factor into the game audio, but there will be times that these changes will be mixed into the sound files before delivery into the game. You can begin to imagine the large number of sound files that can be generated for a game—oftentimes over 20,000 lines of dialogue alone! Now you can begin to appreciate my taking the time to discuss documentation and proper script marking for the proper takes earlier in this book. The naming-convention process is vital to the proper functioning of the game, as well. Attention to detail in this world will test your skills as an engineer, but working in game audio can be very satisfying.

RECORDING FOR ANIMATION

Recording the voice for animation is similar to the process for games, but there can be some significant differences. Unlike with games, we have the luxury of recording more than one actor at a time, which can liberate the performance. It stands to reason that an actor will give a much better performance when she is reacting to another actor in the studio, rather than having to imagine the other side of a conversation. Many times we will have a "reader" in the studio with the actor to deliver the other lines, which gives our actor something to react to; often this will be the director, or the sound or dialogue supervisor, or some other such person. However, actors like to act with other actors, and, when we can get two or more into the studio at the same time, the level of acting increases at a huge rate. An actor in the studio all by herself can become very lonely, and, although the ultimate performance will be fine (after all, these folks take pride in their work), it's simply not the same as having someone else to bounce the lines off, and the experience gives all of the actors a sense of participating in a real conversation and a real experience.

If you are going to try to record multiple talent at the same time, there are a couple of things that you should be cognizant of: First of all, try to keep the two (or more) microphones as isolated as possible, in case you have to intercut another take into the performance. You don't want bleed from the second actor's mic from the original recording to be heard under the new line, and a part of avoiding this is to listen for overlapping dialogue very carefully. Know when you can allow the overlap and when it might become problematic. Second, because things are going to flow naturally with multiple talents in the studio at the same time, it's easy to get away from the scripted lines. You should always record the script as is at least once, and preferably numerous times, so that you have coverage. Then, and only then, allow the actors to begin to improvise. Again, always get the written lines first, and then invite the actors to begin bringing themselves into the role. By doing this, you can save yourself endless headaches in the future, when it comes to editing and client approvals.

Recording dialogue for animation can test your skill at session management, as well: As the actors begin to get into their roles, the energy level often rises very quickly in the studio, and the performers will start to take off into creating this new world. Know when you have to step up and rein the actors in, to keep them on track and to keep the session running along smoothly. Both Jonathan Winters and Robin Williams were famous for cutting loose in the studio, and they would improvise for hours before anyone could get a word in and remind them to read the script! This is not an uncommon occurrence in the world of animation recording; actors love to improvise and try and crack each other up. This can be a lot of fun for everyone connected with the session, but, as the engineer, you have to keep in mind that this is your show, and that the session has to keep moving along at a set pace. Diplomacy is vitally important in this regard—you don't want to come off as being a Scrooge and someone who is killing everyone else's fun, but, at the same time, the session has to finish up sometime before the sun goes down! The longer you let the improvisation continue, the more of the client's money is being spent, so watch the clock and make sure that everything gets accomplished. All too often, in the high-energy atmosphere, lines go missing, and redos are necessary. Keep an eye on that script!

WATCH YOUR LEVELS!

In any type of voice recording that we do, one of the cardinal rules is, "never distort the recording." When recording for games and animation, this can sometimes become very difficult; in the space of a single line, our character can go from a whisper to a scream, and even the best of mixers can't react fast enough to pull down the fader in time to avoid clipping and distortion. A compressor/limiter can help you, but quite often you can hear it working, and that's not a satisfactory solution, either. So, what to do?

If we set our fader and microphone preamp for the loudest sound, the whisper can be lost in the noise floor of the room and the signal chain. Likewise, setting for the quietest means instant distortion if the actor screams. Finding a nice balance, somewhere in-between for the fader or the mic preamp, gives us a good compromise, but keep in mind that you are a human, and you are a mixer. Keep your fingers on the faders and be constantly adjusting; in essence, you are a human limiter! However, there is still a chance of distortion on a very loud line of dialogue, because, no matter how fast you are with a fader adjustment, the mic preamp will be clipped, or possibly the mic itself will distort. The best way out of this is to use two microphones, set up either side by side or with one about 4–6 inches back from the front mic. This second mic is set up as follows: Making sure that the two mics are complementary in sound and capable of being intercut in the edit, but that the secondary mic has a good deal of headroom before the onset of distortion, set the mic preamp 8 dB or so less than the "main" microphone. Now, if an extremely loud line of dialogue catches you off guard, you can cut in the line from the second (or "yell") mic, match EQ and level with the first, and *voilà*—you have avoided the distortion on your primary mic. This technique has been perfected by the folks at Disney and at Pixar, and, if they are using an outside studio to record an actor, this is what they specify for their recordings. Of course, it is possible that the second mic could distort as well if the line is delivered strongly enough, but this extra headroom pad almost always solves the problem.

This dual-microphone technique is very useful if you are recording yourself and you can't pay 100 percent attention to your metering and mixing during the read. It not only can protect you from clipping, but it gives you a nice confidence while recording that your tracks won't be distorted. You can stop worrying about whether you are clipping and instead give more concentration to the performance. You should be aware that it will take some experimentation to find the ideal mic preamp settings, and this experimentation should be done prior to the recording itself. Set the signal chain for a normally spoken script, and then set the second mic and its signal chain while really belting it out. If everything is set correctly, you should have full confidence in the tracks you deliver. Of course, this will take some editing to put it all together, but that should be a part of your delivery anyway.

INSIGHT: Vince Caro, Chief Recording Engineer/Sound Designer, Pixar Animation, Emeryville, CA

It's been my experience that every successful recording session begins with one thing, preparation. It doesn't matter if I'm recording an eighty-five-piece orchestra, a jazz big band, a string quartet, a rock power trio, or a single voice. The one common denominator is preparation.

For me, preparation begins immediately when I'm contracted for the job. I try to gather as many details about the session as early as possible, so I can begin to mentally prepare and plan. Here's part of my mental checklist:

1. In what recording studio will it take place?

   (a) Am I familiar with that room?
   (b) Are the management and staff competent?

(c) Is it technically buttoned up?

(d) Who will my assistant be, and am I confident in their abilities?

2. How many musicians?

3. What is the instrumentation?

4. And this one is always difficult to obtain, but it's always immensely helpful when you can ask the artist or producer, "What are you attempting artistically?"

The reason number 4 is so important is that it's really the best "early opportunity" for the engineer to build into the recording session the best technical solutions that will best facilitate the artist's vision. The engineer can prepare to design the room setup with proper sight lines, points of rejection for the microphones, the choice of microphones, mic preamps, and recording medium. All of those seemingly minute considerations often have immense ramifications later on. For example, if you were asked to record music for a documentary that took place in the 1940s, would you plan to record it using vintage equipment? Would you have all the musicians in the room playing live together? In my experience, almost every decision you make when you *assume* what someone else wants is wrong, so it always best to get as much information as early as possible. But one word of caution, please do so without making yourself an annoying pest! Being flexible and easy to work with is also very important in developing successful relationships.

If I'm ever in that unenviable position of heading into a session not knowing what the artist or producer is looking to achieve, I overdo my setup. I go into "prepare for any request" mode. I'll have my "best guess" setup ready to go, and I'll have other mics and preamps standing by (usually the most strategic, based on which will probably make the most difference). If I don't know if isolation booths will be needed or wanted, I will at least have them cleared and standing by, and I might have some mics set up for what I think might go in there. Of course, I often guess wrong, but sometimes "happy accidents" occur because of this kind of preparedness, or, worst case, a quick and easy change can occur, and we can move on quickly before the inspiration dies. I find that, if I'm at least prepared mentally, and as technically prepared as possible, I can then truly roll with the punches. I have found that, in many cases, true artistic magic has happened because I was prepared for anything, and "*anything*" happened.

I apply that same philosophy of being prepared and ready for any request with regard to recording dialogue at my studio at Pixar Animation. True, I may only have two, four, or eight microphones to worry about, and not thirty or forty, as I might have on a scoring session, but, as is often the case, requests are made at the last minute, and the thing you never want to have happen is to miss a take, or for the "creative" to lose their moment of inspiration waiting for you to set something up, or fix something that's not working. I start recording the moment the talent walks in the door, and I don't stop until they walk out the door. And the only way to ensure that happens is to have the studio set up, tested not twice but three times, and ready to record. The room should be neat, clean, well lit, and inviting for every person who walks in, right down to fresh pads of paper and sharpened pencils. There should be bottles of water for both the acting talent and everyone else in the room, mints, cough drops, throat sprays, and anything you can think of to make the talent—*all* the talent—happy and comfortable. And, of course, the thing that makes everyone the most comfortable is to know that the engineer is not just competent, but totally in control of everything technical in that studio. They need to know that, no matter what happens, they will walk out with a great sounding recording.

I approach dialogue recording the same way I approach music recording with regards to artistic vision (i.e., number 4 from my checklist above). If I've had a chance to talk to the director or editor about "anything particular I should know beforehand," I often find out that they haven't really thought about the "sonics" of what they might want to achieve. But often, once I hear what the scene is about, I can suggest and easily help them design the sound as I am recording it. Truth be told, most often the direction is to get the dialogue recorded clean and totally unaffected, but, every once in a while, there are opportunities to design the raw recorded sound. Here are some examples: For one film, I was told that we would be recording a 1930s or 1940s newsreel-type voiceover. So I broke out my vintage RCA 44 microphone and RCA OP6 mic preamp, which totally nailed that vintage sound. Sometimes, I'm asked to record something that might sound like an NPR interview, or a television commentator. I'll have another microphone set up, maybe an EV RE20, or a Shure SM7, or just quickly put one of those big foam ball pop screens over the main dialogue mic (Neumann U87) and have the talent get right on the mic, "FM-radio-like," and compress and equalize accordingly. For a TV commentator, sometimes a lavalier on the lapel or shotgun above the head is the way to go. There have been times when I've had to mimic the sound of someone with poor mic technique, like someone talking on a podium at school assembly. Break out the SM57 or 58 and have them let the plosives (P pops and Bs) fly! I've also had some great results having talent speak their lines into cardboard boxes, through cardboard tubes, into giant galvanized or plastic garbage pails, or five-gallon water bottles, to achieve certain sounds you just can't get with a plug-in. In all of those "instant sound design" moments, I almost always (when possible) simultaneously record the voice clean. Obviously, when your talent's head is inside a metal garbage pail, that reverberant sound goes everywhere, and so you'll need to do separate clean, unaffected reads, if your director is not sure that it is the exact sound they want. It's always best to cover it both ways.

As far as my normal setup for dialogue recording goes, my main microphone is a Neumann U87. My backup, or "scream mic," is a Brauner VMA. Sometimes, I'll have a Schoeps CMIT5 or another longer shotgun set up, if we're anticipating a more distant POV. I do like the U87 for dialogue; it's an excellent all-around mic and has served me very well over the years. Of course, there are many other excellent microphones that I could use as my main dialogue mic (the Brauner VMA, for example), but practically speaking there are few better choices than the U87, simply because it is the one truly great microphone that is also ubiquitous. Every professional recording studio worth its salt has at least one, and most have a half-dozen or more. And, as Pixar and Disney travel around the world to record the voices for their animated features, consistency is a major factor. All Disney and Pixar animated features take about four to six years to make, and you have to have consistent-sounding dialogue. It's not unheard of to have part of a line living next to another part that was recorded four years earlier or later, and every point in-between. So, not creating headaches for the editors and the mixer doing the final film mix is of the utmost importance.

There have been occasions when I used a microphone other than a U87 to record certain voices in an animated feature. For example, I used a Telefunken 251 to record Paige O'Hara as Belle, in *Beauty and the Beast*. That choice was a carryover from the soundtrack/song recordings. Her voice sounded wonderful while singing into that 251, so we used it for her speaking lines as well. For James Earl Jones as Mufasa in *The Lion King*, I used a vintage RCA KU3A. I believe Doc Kane, the other dialogue recording engineer/ADR mixer who I worked with on this film, made that suggestion. But these kinds of decision are not made lightly. As I said before, consistency is very important. In both of those cases, I maintained the availability of those mics for the entire project.

The next item in the recording chain is the mic preamp. Again, we also like to keep a certain amount of consistency for our standard dialogue recordings. I prefer the Focusrite Red 7 for both our U87 and VMA. I always travel with my pair of 7s, which are from the earliest manufacturing run and were built with those huge Lundahl transformers that sound so "fat." I think I'm responsible for about a hundred other studios around the world buying them, after they'd heard my work with them. They are standard in the recording studios here at Pixar. The Focusrite Red 7 offers tons of headroom and has some built-in essentials, such as a lovely high-pass filter, for studios with pesky air-conditioning rumble, and a very flexible compressor. I try to not compress our dialogue, and so I set a very high threshold that is there simply as a safety net. For recording screams and super-loud lines, I almost always take the compressor out of the chain and go "commando," relying upon my skill not to allow distortion. Few things sound worse to my ears than a compressed scream; I prefer "unaffected amplitude"—it just sounds better—but you have to be good to get it right!

For the unexpected scream or yell, which almost invariably saturates the microphone capsule, I always have the aforementioned scream mic, the trusty Brauner VMA. For many years, my scream mic was a Neumann TLM170, set to hyper-cardioid.

As I recall, the TLM170r has about 17 dB more headroom at the capsule than a standard U87 Ai, and so it fulfilled that job perfectly over the years—I never lost a take! But, simply put, the Brauner VMA just sounds more open, and it has about the same headroom as the TLM170, or more. That said, few studios have Brauners (they are very expensive), and so we still rely upon the TLM170 for most recordings outside Pixar's or Disney's studios. Another bit of information is that I always set my scream mic about 4 inches further back from the talent than the U87. That added distance and the added headroom of the Brauner or TLM have aided me with those unexpected screams, or in the odd happenstance when the director has asked for a whisper, and the talent screams. It happens, and there's not much you can do, which is why preparedness is key.

As for pop screens, I prefer the Stedman PS101. It's a plastic-covered metal screen that is super-transparent compared with the foam ball or the more standard "pantyhose" pop screens, which often give you a tearing sound when air passes through. I did make one modification to my original Stedman pop screens, and Mike Stedman has since adopted this modification for some of his newer models. I found that, when they were hit with just the right level and frequency, they would "ring." So, I applied a rubber tubing grommet to the outer edge of the screen, which solved the problem. I alerted Mike to this issue, and he has since modified his welding technique and now offers the option of his screen with a rubber outer ring. The other benefit to the Stedman is that we often record video of our talent, so that the animators have the option to observe facial expressions that the actors might make when they are acting. Other pop screens block the face too much, but the Stedman is also the most transparent visually—a real bonus!

My preferred A/D converters are made by Daniel Weiss, but I also like the Lynx Aurora. They both have been reliable and sound great. Currently, we record to Pro Tools, and I run a separate backup recorder in the event there is a crash. That backup machine is a Fostex DV824, which can record to both an internal hard drive and a rewritable DVD. It can record up to eight tracks of up to 24 bit/96 kHz. It simply shows up as an additional hard drive over the network. It's a handy little machine!

AN EXPERIMENT IN VOICE RECORDING

Although the vast majority of voice recording for animation work is done in the controlled environment of the studio, it's worth noting that there have been some interesting experiments with recording voice lines for feature-film animation on location, as if the film were shot live-action. This manner of recording lies somewhat outside the scope of this book, but it is worthwhile to discuss this technique briefly.

In 2011, the director Gore Verbinski released the animated feature film, *Rango*. For this project, he wanted a more natural performance from his cast, and so the decision was made to get out of the studio and allow for a more naturalistic experience for those involved. To this end, the dialogue was recorded using the techniques normally associated with live-action filming—all of the cast interacting together, instead of one or two actors in the studio at a time, and having the ability to move around and physically act out a scene. He assembled the cast on a sound stage, with a limited number of props and costumes, and, after a good deal of rehearsal, began, not only recording the script, but also filming the actors as they performed. Character actor Abigail Breslin is quoted as observing:

> *We filmed it like a play and we were all together. If an actor you're working with changes the way they do something, you tend to change the way you do something. It's a lot more fun than standing in a recording studio, talking into a microphone.*[1]

The dialogue was captured using standard film production equipment (shotgun microphones on boom poles and miniature lavalier microphones), and this allowed the actors freedom of movement during the scenes. If one of the characters was supposed to be riding a horse, they would sit on a saddle mounted on a wooden stand. If gunshots were called for during a scene, all of the actors were provided with earplugs to protect their hearing, as blanks were fired on set. The entire production crew raved about the resulting performances, and the actors were convinced that this method of recording for an animated feature was the most liberating experience they had had when doing animated voices.

It should be noted that *Rango* was not the first animated feature to be captured in this way. Wes Anderson used a similar technique in creating *Fantastic Mr. Fox* (2009), and the technique (as is so often the case in animation) was first used by Walt Disney with *Alice in Wonderland* (1951) and *Peter Pan* (1953). The combination of live-action performance capture and an animated story line can result in a powerful end performance, and, if you have the resources to experiment with the technique, it is well worth the effort.

The aesthetic philosophy behind these experiments can be summarized as the filmmakers wanting to capture the ambience and perspective in a natural way and not to try and emulate these in post-production. Can these techniques be brought to other types of voice recording—commercials, for instance? In the late 1980s, I did experiment with recording a few radio commercials on location, if the script demanded an external scene, and the results were mixed. On the one hand, the ambient sounds were very convincing, and there was no need to recreate them in the studio using sound effects. On the other, intercutting various takes to achieve a good final mix

---

1 KRCR News: krcr.com/programming/entertainment/ at-the-movies/making-rango-a-group-effort-for-stars/115670

was problematic, at best. I do believe that further experimentation along these lines would prove fruitful, and I hope that some of you attempt this technique and pass your findings along to others in the industry. At present, I am unaware of this technique being used for game audio, but, for certain "cut scenes," it might be useful for gaining a feel of the real world and convincing perspective. Again, further experimentation should be undertaken.

Recording Interviews and Roundtable Discussions

12

Although not strictly voiceover recording, an area that we should explore and one that you may find yourself asked to do, is the recording of interviews and roundtable discussions. This is a skill that lies somewhere in the middle of studio voiceover recording and recording dialogue for film and video on location (known as "production recording" or "location recording"). Production recording is outside the scope of this book, and there are many fine sources of information on the subject; however, it is something that you may want to explore, as it gives you a new skill and another chance to become employable and have another income stream, which, after all, is what we all strive for. For recording interviews in the field, we often use the same equipment that is used for production recording and many of the same skills; therefore, we will take a look at the gear and techniques that are essential to recording in the field.

There are times when you will be asked to record an interview that is conducted by someone else, and there might be times when you will be the interviewer as well as the person recording the interview. This chapter will address both of these situations, as well as strategies to effectively record roundtable-type discussions.

Interviews are conducted in a wide range of environments, one of which is the studio setting that we are all used to and comfortable in. Often, however, we have to get out of the studio and go into the wider world to do our work. Recording on the street, in a shopping mall, grocery store, farm field, sporting event, office, conference room, home, airport, or other unlikely place certainly will test your knowledge of microphones and acoustics, as well as your ability to think on your feet and improvise. Also, you will have to have an intimate understanding of portable recording equipment: If you are fumbling with the controls and having problems getting things up and running in a short period of time, your subject will never give the best that they are capable of. When it comes to the gear, you should be able to operate it in the dark. If you are interviewing the "man on the street," the more you are unsure of what you are doing, the more nervous your subject will become, and you will not capture the best possible product for your client.

Now, having said that, I want to tell you a little story about one of the best interviewers that the world has ever known. His name was Studs Terkel, and he was one of the most decorated writers of the twentieth century, winning the Pulitzer Prize in 1985 for his book, *The Good War: An Oral History of World War II*. His work was, for the most part, oral histories on various topics, such as *Hard Times: An Oral History of the Great Depression* (1970) and *Working* (1974). Although he had interviewed literally thousands of people over the years, he had a trick that he pulled time and time again: When sitting down with a subject and putting his tape recorder on the table, he would pretend that he didn't understand technology and he could never figure out how to work the machine. The interview subject was immediately put at ease, because, for the most part, they don't understand recording technology either, and before they knew it they were deep in conversation with Studs and didn't even realize the tape was rolling. So, figuring out how to put a subject at ease is one of the little tricks that will make or break the interview, and one that you will have to learn with experience. No one technique is going to work for everyone, and so being able to read people is of the utmost importance for this type of work.

Remember, everything that we talked about in the first part of this book still applies, and you should be aware of acoustics, signal flow, documentation, and so on. You might want to review the chapter on microphone types and polar patterns and start thinking now about what type of microphone you might consider using on location, instead of a large-diaphragm condenser that we would use in the controlled environment of the studio. The recording of interviews is an exercise in overcoming conditions that are not ideal, and so, the more knowledge you can bring to the job, the better off you will be, and you will arrive at a surprisingly high-level, consistent product.

INTERVIEWS

Unless you're conducting the interview in a studio, the location sound for the interview is going to be unpredictable. For instance, if it's on the street, the time of day it is being conducted will have an impact on the amount of traffic noise that you'll be contending with. So, one of the first things that must be done when you arrive at the location is *listening*. Of anything you do in audio, nothing is as important as to just stand still and listen, and doing location recording of any kind is no exception. Then, of course, you must decide what to do with what you hear. Let's go back to the example of recording on the street: If you are using any kind of directional microphone, is there an advantage in facing one direction or another to help minimize the traffic noise? Is there a better time of day to conduct the interview, because of the schedule of airplanes overhead or trains passing by? If the interview is taking place indoors, are there noisemakers that you can quiet? Let's say you're recording in a home, and the noise from the refrigerator is bothering you. Well, how about turning off the refrigerator for the duration of the interview? (Hint: If you are going to take this route, put your car keys in the fridge. That way, you remember to turn it back on before you can possibly leave.) If there is possibly anything you can do to lessen or eliminate background noise, do it. Recording in an office, and canned music is an issue? See if you can get it turned off for the duration of the interview. If not, find another location—background music will make it impossible to edit the interview into anything coherent without having disturbing jumps in the music. Being aware of any noise source prior to turning on your recorder will help to make your interview a success.

The reasons for conducting interviews are as varied as you can possibly imagine. Interviews can be used as on-air testimonials for commercials or in podcasts; they can be transcribed for use in print advertising, for web content, corporate training, and more. All of these are reasons to conduct interviews, and each has its own problems and limitations. Of course, you should ask the client if the interview is to be audio only or on camera; the answer to this question may drive your choice of microphone and method of working. All of these questions will impact the way in which you approach the interview and the equipment you might choose to do the job.

Let's start this discussion with the assumption that you will be both interviewing and recording your subject. Obviously, you want to keep the amount of equipment to a minimum and be fairly lightweight and portable; lugging around a heavy recorder all day can really tire you out. Usually, in an interview situation, you will be talking to only one person at a time, and so the need for multichannel recorders and elaborate mixers won't be an issue. There are many excellent portable mixers and recorders made today, and they get more compact every year. Recorders are available in configurations from two-track stereo to eight channels, and so every location possibility is covered. I've had very good success using a stereo audiocassette recorder and a quality microphone in years past, as well as the latest generation of handheld digital recorders. You're limited only by your budget and what you're trying to achieve. At present, one of the very good digital recorders is made by Zoom and includes built-in stereo condenser microphones that are quite good, as well as the ability to record up to six independent channels using mics of your choice. These recorders allow you to choose your file format (.wav, .aiff, MP3) and sample and bit depth, and many come with expandable memory in the form of SD cards. The Zoom H4n is one of the most used pieces of gear for collecting sounds on the market today, and recently the H6 was released, featuring interchangeable capsules and giving a wide range of options, including a shotgun microphone attachment and the ability to record up to six channels of audio from external microphones. As well, there are digital recorders from Sony, Tascam, Olympus, Nagra, and many more; again, your choice will depend on features and your budget. Lightweight and fairly rugged, these recorders are a great option for an engineer on the move and can be put into the pocket of a jacket or clipped onto a belt to keep them out of the way. The recorders have quality microphone preamps built in and do a surprisingly good job, considering the size and price point.

Not too long ago, the recorder of choice was the DAT recorder, and it did a very good job. The format was short-lived, however, acting as a bridge between the analog days and the coming of full digital, and the tapes were expensive and not very robust. Also, in years past, a number of people got very good results using the

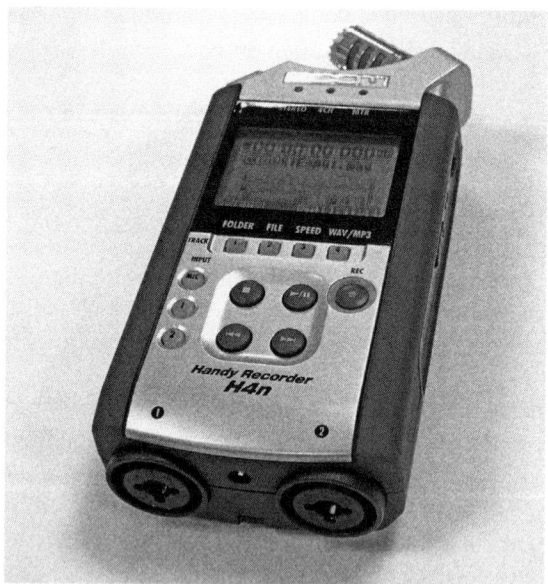

FIGURE 12.1

Zoom H4n Handy Recorder

Source: Photo by the author

mini-disk recorder, but these have also faded in popularity as the handheld digital recorders came on the scene. The handheld recorders have the advantage of connecting to your computer via USB cable and appearing as an external drive to the computer; you simply drag and drop the files onto the computer, import them into digital editing software, and you're ready to edit and mix. They have a surprising amount of memory (expandable in some models) for the price, so that long sessions and backups are not an issue.

Of course, you can also record to a laptop using software such as Pro Tools, Logic, or any number of other programs, but you're forced to carry around both the computer and an external interface to a good-quality mic preamp and monitoring, so I don't consider this a very good solution for recording in the field. Also, I wouldn't recommend carrying your laptop around if it's raining, snowy, or dusty, and so this isn't a good option. However, as we begin talking about doing roundtable discussions (as these are usually done indoors), we'll see that the laptop is a good alternative for recording purposes.

Other options include using one of the many types of portable field recorders and mixers on the market, from manufacturers such as Sound Devices, Zaxcom, Sony, Tascam, and others. Intended for the film and television sound industry, they were designed with the expectation of their being used outdoors, in a wide range of conditions. A quick web search will reveal a large variety of products, in various price ranges and with various features, available in models that include time code (if you're recording to picture) and with a choice of number of inputs. One of the

FIGURE 12.2

Sound Devices 744t

Source: Courtesy of Strapko Recorders; photo by the author

more common to find in the field would be the Sound Devices 744, with four inputs and a full range of features, including 48-V phantom power for condenser mics, a bass rolloff switch, limiters, and independent control of input levels (for time code, use the 744t). Again, like the handheld units, these models allow for a choice of audio file formats and sample rates to match any requirement.

As with anything audio, you should carefully consider your needs and budget and try a number of units to make sure which is right for you. And, as always, purchase equipment with an eye to building your personal collection for the long term. You want to be able to live with your decision for quite a while.

Before we take a look at some microphones that are commonly used for this type of work, I should remind you that mic choices are highly personal, and we all have our own preferences and needs. What I might think sounds wonderful in a given application may not be what you want to hear, and vice versa, so once again I will point out the need to experiment and carefully listen and evaluate, whatever equipment you are considering.

As for the microphones commonly used for recording interviews, much of the decision will depend on whether you are recording audio only, or if the interview is being filmed. Doing a one-on-one interview most often is done using a handheld mic; it's easy to handle, highly portable, and can be used to prompt the subject to complete a thought or give a more in-depth answer. Whatever your choice of microphone, always have a windscreen on the mic: Many times, the interview will be done outdoors, and wind noise creates an unacceptable sound that you must avoid. If you are recording indoors, the windscreen will minimize popping on the mic, another distraction that you shouldn't allow. If you are the one conducting the interview, you'll probably want to avoid wearing headphones, because you want to have a conversation with the subject, and most people feel a little odd speaking with someone who is both pointing a microphone at their mouth and isolating himself with headphones, and so it won't be apparent if your mic is popping; the use of a good windscreen is a very effective insurance policy.

You'll also want to minimize handling sounds from the mic body; many microphones are very susceptible to handling noise. In general, dynamic mics are better at minimizing this than condensers, but you should both experiment with the mic before purchase to test for handling noise and also practice using the mic quietly and smoothly. Like anything else, there is a skill involved. One way to avoid handling noise is to use a shock mount, such as a pistol-grip mount with a shock mount as part of its design. They are very effective, but they force you to point the mic ahead of you, rather than up at the interview subject's mouth, and trying to point it at your own mouth to record the question is very problematic.

For most interview situations in the field, I prefer a dynamic microphone, for a number of reasons: As stated, they inherently have less handling noise; they are generally more rugged than a condenser and can take traveling better; they don't require phantom power; and they are less expensive to purchase in most instances. This last point shouldn't be overlooked. If you happen to get caught in a rainstorm, or your equipment bag gets misplaced or stolen, you're out less, and your favorite condenser studio mic stays safe and dry. There are some very good-sounding dynamic mics on the market, and more are coming along almost every day it seems. One of my favorites for doing both interviews and roundtables is the Beyerdynamic M201. It's a hyper-cardioid pattern and is very smooth and clear sounding, without much artificial boost across its frequency spectrum.

The mic has a very low profile, is comfortable to hold, and functions well as a general-purpose mic for both speech and instruments. It's also surprisingly affordable for such a rugged little mic.

Some people prefer a microphone with an elongated body, because they are easy to maneuver, and you can get very close to the subject's mouth. One of these would be the Shure SM63L (the "L" in the model number designates the long body style). This mic is omnidirectional and was specifically designed for interviews and newsgathering. As with most of the Shure dynamic line, it is extremely rugged and dependable, but I find the sound a bit on the thin side. You decide for yourself. Many people prefer the omnidirectional pattern for interview work, in case the subject suddenly turns his head; with a cardioid or hyper-cardioid, the voice would dip in volume and produce an off-mic sound as the subject moves off axis. The omnidirectional mic will help with both these and increase your chance of success. It also allows for recording both the interviewer and subject without as much mic movement, for both the questions and answers. I would recommend doing some online searches for various microphones and reviews, as well as talking to a representative from any pro audio retailer for help in choosing a mic.

A good, all-purpose microphone for voice work is the Sennheiser MKH416 that we mentioned in the chapter on microphone types (see Figure 4.19). This short shotgun mic is highly versatile and can be used in a wide range of circumstances.

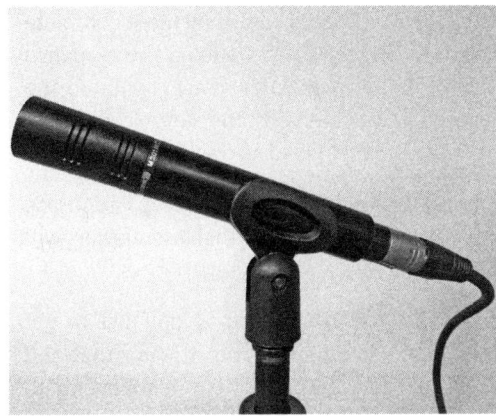

FIGURE 12.3

Beyerdynamic M201

Source: Courtesy of Strapko Recorders; photo by the author

FIGURE 12.4

Shure SM63L

Source: Courtesy of Shure, Inc.

INSIGHT: Ray Van Steen, Interviewer and Voice Actor, Dallas, TX

I got into interviewing kind of by accident: It started with me being asked to do man-on-the-street TV interview commercials. One day I got a call to ad lib some television interview spots for Camay. The camera was behind me, shooting over my shoulder, filming the comments as I questioned a number of young women who used Camay. It was tougher than I thought it would be. I discovered that merely asking memorized questions didn't work. An actual *conversation* had to take place. I had to ask questions that couldn't be answered "yes" and "no." It was a matter of getting them talking.

Interviewing real people for use on radio or TV testimonial commercials is completely different from the interviews that talk-show hosts do with their guests, because, on commercials, the question part of the conversation is almost never heard, and the interviewer is rarely seen. The answers provided by the interviewee must be able to stand completely alone. Don't pose questions that can be answered "yes" or "no."

One of the best ways to avoid getting a "yes" or "no" answer is to never actually ask a question. For example, don't ask, "Do you like the taste of this new Betty's Butterscotch Pudding?" Instead, say, "Tell me about the taste of this pudding." "Tell me about" or "Talk to me about" are great ways to elicit statements that are not simply "yes" or "no."

The interviewee may reply, "It's great." Of course, we need something more complete than "It's great." The interviewer then says, "*What's* great?" Usually, the response comes back, "Betty's Butterscotch Pudding." Well, *that* can't stand alone, so the interviewer then says, "Put that all together for me." Ideally, the interviewee then comes up with what's desired: "Betty's Butterscotch Pudding tastes great."

This isn't prompting or leading, it's simply asking the interviewee to rephrase what's already been stated. After a few go-rounds like this, the interviewee will have "learned" to answer in complete statements.

Of course, sometimes it's necessary to ask a "yes/no" type question, just to get the interviewee talking, or to introduce a topic. If the situation is such that you must ask a question that requires a "yes" or "no" response, use a follow-up to get a more complete statement. For example:

> "Do you like the taste of . . .?"
> "Yes."
> "Expand on that for me."

Or,

"Really? Tell me why."

There are legal and ethical considerations when using real people in testimonial commercials. The interviewer must take care not to put words in their mouths, nor to lead them. There is a difference between "prompting" and "leading." Let me give you a good example: If I ask, "Does it taste good?", there are only two possible answers: "Yes" and "No." Very dull material for a commercial. But, if I say, "Tell me about the taste," a wide range of comments is possible. I never lead anyone. Being aware of the tricky legal ramifications in testimonial advertising is important.

An inexperienced interviewer could make a mistake in the middle of an interview that could completely disqualify, on legal grounds, an otherwise great testimonial. Legal opinions vary on what actually comprises leading, or prompting, but I've found that there's no problem in suggesting areas for an interviewee to explore. Here are a few ways to do this:

"We've had some people tell us that Betty's Butterscotch Pudding is creamier than some others. What do you think?" Or, "Y'know, the folks at Betty's say that they've gone an extra step in making their pudding creamier." Just let the statement—not a question—hang there as you look expectantly at your subject. Most people will make a comment.

There are two main difficulties that you run into when interviewing "real people": answers that are too short, and those that wander on and on. The exact opposite of answers that are too terse are answers that ramble. Some folks just have trouble packaging their thoughts into short, pointed statements, which is what's needed for use in testimonials. Following such a rambling answer, the interviewer might say, "Gee, that's a lot of great info, but can you put all that in a shorter statement?" When interviewees are asked to repeat in this manner, they tend to leave out needless details. If they don't, they can be told specifically. "Okay, can you tell me that again, but leave out the part about what time of day it was, and don't go into all that detail about the people who were present when you first tasted Betty's Pudding."

Another of the difficulties encountered is extraneous noise that you know will ruin an otherwise great comment. For example, when doing man-on-the-street interviews, there are often horns honking, brakes squealing, etc. When something like this happens during a usable interviewee statement, I usually roll my eyes in reaction to the noise, and say, "Wow, that was a great thought, but the horn honked just as you were saying it. Could you repeat that remark about telling your neighbor about Betty's Pudding?" Be aware of the "liftable" comments and be aware of other sound sources during the interview. I sort of edit in my mind. If I hear an absolutely beautiful comment, I instantly think of it as "liftable" or not. Maybe there was a car horn in the background, or perhaps the person slurred a word or something. If the remark isn't perfectly liftable, I try to get it repeated. I pretend I didn't hear correctly by saying, "Pardon?" Or I'll wait until later in the interview, and try to steer the person back to the same subject area. Also, can the statement stand alone, without an explanation from an announcer? I like the trick of pretending that you didn't hear something, so that the subject repeats the statement. "What was that? Could you repeat that, please?"

Being brought into the process at an early stage certainly helps the overall product. Sometimes, I'm the *first* person the ad agency calls. They tell me what they want and turn me loose. I grab my tape recorder and take care of virtually everything, including supervising the editing and production of the finished audio tracks. For most television projects, I'm usually called in by the film company that has been hired by the ad agency to produce the commercials. In most of those cases, I function just as the interviewer. The agency or film-company people handle the logistics, selection of people to be interviewed, and other details. Sometimes, early in the planning, the people I work for will hand me a list of questions that they want me to ask. I hang onto them for reference, but I always tell them that I'd much rather have a list of the kinds of *answer* or *statement* they want from the interviewees. Based on that, I can figure out the best way to elicit them.

It is a condenser and it is a hyper-cardioid, but its usefulness in a range of applications makes it a good mic to have in your collection.

Besides the equipment used for the one-on-one interview, there is also the technique of conducting the interview. By far, the number one thing to keep in mind is that you want to have a *conversation* with the subject, and not just read off a list of questions.

VIDEO INTERVIEWS

To this point we have been discussing one-on-one interviews, with a handheld microphone, for audio-only recording. Video or film interviews are a bit different, and the fact that there is a camera rolling may change your choice of equipment. Even if you are doing an audio-only recording, but someone else is doing the interviewing, your equipment will probably change. As for the recorder, not much changes, with two possible exceptions:

First, you might want to send your audio directly to the camera. That way, all of the media remain in one place (on the videotape or card that the camera is recording to), and you can't get caught in the mismatched-time-code scenario. If the video and the audio are at differing frame rates, things will never sync up properly, and all kinds of problem will arise. Recording directly to the camera is called "single system" recording, because there is only one recording device.

Second, you could use a stand-alone recorder (this is known as "dual system" recording). In this case, you should seriously consider a recorder that has time-code availability. Although it is certainly possible to record without the use of time code (with the digital recorders—this would never work with an analog tape recorder, owing to speed fluctuations inherent in analog equipment), syncing and editing are greatly aided if everything has matching code. In either case, you should do both an aural and visual sync match at the beginning of the take, either with the use of a slate (clapper) or by standing in front of the camera while both it and the audio recorder are recording and clapping your hands together. This will give both the video editor and the audio editor a mark to sync to, rather than trying to match mouth movements to sounds. The handclap only lasts one video frame, so the sync will be much tighter.

BOOM AND SHOTGUN MIC TECHNIQUES

There are times when you will be recording someone else who is doing the interviewing, either for an audio-only interview or as part of a film crew that is taping the interview for future use. In this case, you would most often use a shotgun mic on a boom pole or a lavalier mic placed on the interviewee. These techniques are most often not used in the studio (although there are exceptions, of course), and you have to be aware of the different microphone types and their uses. The shotgun mic is a hyper-cardioid-patterned mic, which means that it's highly directional. This is great for helping to reduce unwanted background sounds, but this also means that you have to pay close attention to keeping it pointed directly at the subject's mouth: Even a slight variation is apparent and is very difficult to fix in editing. The same thing can happen if the subject suddenly turns his or her head, and so part of your job in this instance is paying very close attention to proper positioning of the microphone.

Shotgun microphones come in varying lengths; the longer the mic body, the greater the "reach" the mic will allow. A short, 8-inch microphone has a working distance of between 3 and about 6–8 feet from the subject. Obviously, closer is better, because you will be capturing more direct sound and will avoid unwanted background sounds and ambiences, but, if you get too close, the mic will be overdriven and will distort. The microphones are designed with a number of slots on the body; these are known as "rejection" ports and are what allow the mic to have its highly directional character. The longer the microphone body, the more rejection ports, and this allows for a longer "reach," or workable distance. Shotgun mics should never be handheld, because they are extremely sensitive to handling sounds. Always place a shotgun mic on a shock-mount system.

A shotgun mounted on a boom is a versatile tool and allows for freedom of movement for the subject (as would be the case in recording dialogue for a film or television production), but there are certain skills that have to be practiced. When

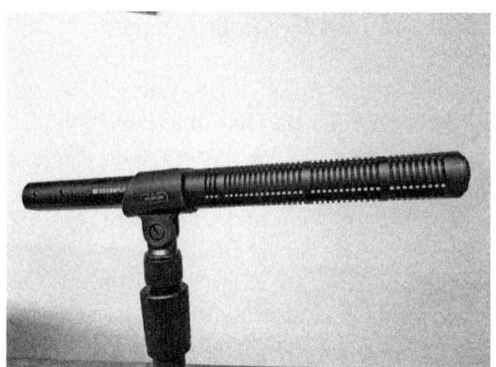

FIGURE 12.5

Sennheiser MKH60

Source: Courtesy of Tribeca Flashpoint Media AA; photo by the author

FIGURE 12.6

Rycote Windjammer Windscreen (Softie)

Source: Courtesy of Strapko Recorders; photo by the author

it comes to holding the boom, make sure that you are coming at the subject from above, and not holding the boom pole in a position where the mic is pointed up at the person. If you come from below, you will be getting much more chest sound, which is boomy and very bass heavy. Also, by pointing the mic upwards, you will be capturing more ambience; after all, where is the noise in our world? Point a shotgun mic up and you will pick up more sounds of wind, birds, airplanes, and so on. Now point it at the ground: Not much down there, is there? Also, by coming at the subject from above, you can easily rotate the pole in your hand to capture both the interviewer and the interviewee effortlessly.

When using a shotgun mic on a boom, wind protection is extremely important. A shotgun mic is highly susceptible to wind sound, and a foam windscreen won't be adequate to protect the mic. There are a couple of types of windscreen available that work well: One is the furry-looking "softie."

This product is made from a synthetic material and can provide up to 20 dB of wind-noise protection. The mic slips up inside the windscreen and is firmly attached to the shock-mount system. If you are using a product such as this, care must be taken not to get it wet, because the "fur" will matt down, reducing the amount of noise suppression. Also, be careful when working around lights: If you get too close, the synthetic material will melt, resulting in permanent damage to the windscreen.

There is also a hard-shell windscreen that is very effective. This has the advantage of not matting if it gets a little wet, and these products can offer up to 30 dB of noise reduction. Here, the windscreen attaches directly to the shock mount, the mic is placed inside in another shock mount, and then a back cover is placed on, so that the mic is completely surrounded by the windscreen. These windscreens come in a variety of sizes, so that you can use longer or shorter shotgun mics, as the situation dictates.

FIGURE 12.7

Rycote Windshield Windscreen

Source: Courtesy of Tribeca Flashpoint Media AA; photo by the author

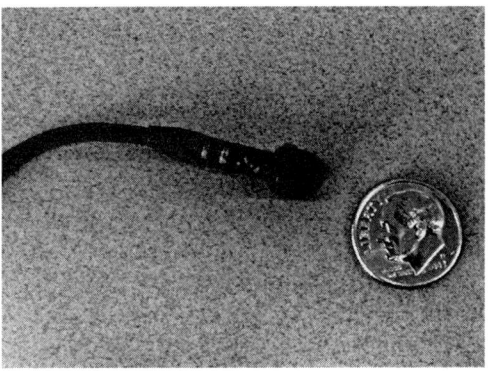

FIGURE 12.8

Sanken COS11xBP Lavalier Microphone

Source: Courtesy of Tribeca Flashpoint Media AA; photo by the author

One other caution when using a shotgun and boom pole combination: Whether the microphone cable runs from the mic and is held in the hand outside the boom pole, or the pole is the type with internal coiled mic cable, be very aware that you can get noise coming from the cable hitting the boom pole, no matter if it is internal or external. When making moves with the boom pole, your move should be smooth and slow, to reduce the possibility of the cable hitting the pole and creating unwanted noise.

To this point, we have considered the handheld microphone and the shotgun mounted on a fishpole. There is another option that is often used, and that is the lavalier microphone. Lavalier mics are small-bodied mics that either operate wired to the recording device or can be part of a wireless system. These mics are often omnidirectional, and care should be taken to adequately shield the mics from wind noise, to which they are highly susceptible. Another concern with the use of lavalier microphones is the sound of clothing rustle. Because they are mounted on the body of the talent and are often hidden in the clothing to avoid detection by the camera, this possibility should be investigated thoroughly, and adequate measures should be taken to avoid this problem. A search of online materials will yield a number of videos that show proper mounting techniques to avoid clothing rustle. Also, be aware that the fabric on which you are mounting the mic will make a huge difference in the amount of clothing noise created. By all means, avoid polyester, nylon, and rough cotton.

As stated at the beginning of this chapter, this isn't meant to be a comprehensive guide to location audio equipment and techniques; there are a number of excellent tutorials online, as well as many books on this subject that go into much greater detail, and, if you find yourself doing this type of work, these are well worth the time to explore.

RECORDING ROUNDTABLE DISCUSSIONS

Recording roundtable-type discussions and interviews, as well as panel discussions, will stretch your abilities in ways that one-on-one interviews don't. With roundtables, we have two main problems to overcome: keeping all of the participants on axis in relation to their microphones and avoiding excessive noise from having multiple microphones open at the same time. For example, say we have four people at a table—three guests and one moderator. Usually, two people in conversation will be pretty routine—two mics, and we're all set—but, as soon as there are three or more participants, things begin to get more difficult. By all means, try to avoid really large groups (I'm thinking of six or more people). In this case, the added noise floor can get very problematic, and the chances of phase cancellation become much higher. For this situation, thought should be given to a special type of microphone, known as the *boundary* mic, which is discussed below. So, for these reasons, let's assume four people are to be recorded.

The first problem that we encounter in this situation is keeping everyone on axis when they are speaking. If there are only two people, situate them directly across from each other, so that they are looking straight at each other, with a microphone in front of each of them. When they speak, they will be speaking directly into the mic. However, with the example of four people, there will be someone who is sitting to the side of each participant. If we position the mic so that our subject directly faces the moderator, she will be on mic when addressing the moderator, but, if she wants to make a point to another panel member, her head will turn, and she will go off axis. I've battled this situation for years and have tried a number of possible solutions: I have used lavalier microphones on the participants, but this only exaggerates the problem, even though a lavalier microphone is usually omnidirectional. The change in perspective is quite noticeable, as is a shift in EQ as the head turns. With a body-mounted mic, there is always the chance of fabric noise as the mic rubs on the shirt or coat when the participant moves about, and so proper positioning is paramount in this case. Another problem with using a lavalier in this setting is that the participants all too often forget that they're wearing it and will stand up at the end of the session, and either the mic ends up on the floor or it pulls the person back to the table. Most lavalier mics have the option of either being hard-wired with the mic cable or running wireless—be cognizant of radio-frequency interference if you're running a wireless system; it can cause big problems, especially in an urban setting.

One thing that you should be aware of when recording roundtable discussions and interviews is phase cancellation owing to a number of microphones being live at the same time and the effect of direct to reflected sound arriving at any one microphone from various surfaces. When speakers are seated at a table, the mic not only picks up a direct sound from the speaker, but also a slightly delayed reflection from the tabletop, and then, slightly later still, a reflection from the walls, floor, and ceiling of the room. One of the best strategies that you can employ is to deaden the early reflection from the tabletop as best you can. Certainly, any kind of cloth covering will help (and will help muffle the sounds of pens being set down, papers being shuffled, coffee cups being moved, and so on), but even more sound absorption can easily be achieved by using carpet, instead of a tablecloth, or better yet, a table pad like many people use on their dining-room tables. Any kind of absorption that you can introduce to avoid the early reflections, as well as muffle any sounds of humans moving, will help you arrive at a clean, usable recording.

The problem of phase cancellation and sound colorization is one of the areas that we have to spend some time solving when doing roundtables. As I just mentioned, the time differential between direct sound and reflected sound can cause a very unnatural sound in the recording, and avoiding this can present quite a challenge. One option that I've had good results with in certain situations is the *boundary* or *pressure-zone* microphone (PZM). The design of this specialty mic produces a smooth pickup, regardless of the source distance from the mic—no off-mic sound, just a change in level. Integral to the microphone's design is a fixed plate, above which the microphone capsule sits at a predetermined and fixed distance. Because sound strikes the mounting plate and then reflects back into the microphone capsule, the distance of reflected sound remains the same in all circumstances, and there is no colorization of the sound due to phase cancellation. Because of this design, the mic only picks up reflected sound from a set distance, and so there can be no off-axis sound in the recording. Although, in theory, this sounds like a wonderful solution, I've found that the sound of a well-positioned condenser microphone is preferable, if at all possible. However, for those times when we are faced with a large number of people at the table, or the requirement for a very unobtrusive microphone presence, the boundary mic can be an option worth investigating.

FIGURE 12.9

Crown Soundgrabber PZM

Source: Courtesy of Tribeca Flashpoint Media AA; photo by the author

This type of microphone is available in cardioid, hyper-cardioid, omni, and stereo configurations. Although not ideal for critical recording situations, where the discussion or interview will be used for final release, I've had very good luck with these mics in a variety of situations, such as a large number of people sitting around the table or in the room, where individual microphones would be impractical. Also, I have participated in recording focus groups for a number of products, where the recording equipment (both video and audio) should be as invisible as possible, to bring out honest comments from the participants. In this case, I would place a couple of PZMs on the walls, ceiling, or floor, as the occasion warranted, and the participants were not aware that audio recording equipment was present. The comments could then be clearly heard and transcribed as needed.

Using a table-mounted omnidirectional mic helps with some of these problems, but I feel that it picks up too much of the room, as well as other participants' speech. A cardioid is a nice balance and a good compromise. Earlier in this chapter, I discussed the Beyerdynamic M201, and I have had good success with that mic in this setting, even though it is a hyper-cardioid. I would recommend this only for those participants who are knowledgeable and experienced in the process of the multiperson interview or discussion, owing to the extremely tight pickup pattern of the microphone. In most instances, I recommend a cardioid pattern on a large-diaphragm condenser mic for these situations. Because the roundtable is taking place in a controlled environment, instead of on a street or some other random place (such as would be the case with the one-on-one interview), a good-quality studio microphone would be appropriate in this situation and yields superior results. And don't forget to use pop guards or windscreens to keep those nasty pops at bay!

Another problem often encountered in recording roundtable discussions is one of a participant striking the table as he makes a point. This is a very natural thing that most people do, but it can be extremely distracting when one listens to the recording. Along with those windscreens, make sure that the mics have proper and adequate shock mounting and isolation.

In this type of discussion, the participants will often have notes with them, and so paper rustle can be an issue. Here is a chance for you to practice your diplomacy to constantly remind them to be very careful when moving paper (a little more on this in a minute, as we talk about mixing strategies for roundtables). Also, if you have a clear line of sight into the studio and can capture the attention of either the offending individual or the moderator, you can use hand signals to warn of the noise problem that you're having. If you are in a studio situation when recording roundtable discussions, you should consider setting up a cue feed to the moderator, so that you can use the cue as a communications link to quietly mention the problem, and the moderator can signal the offender. Setting up a cue feed for communication purposes is always a good idea when doing this type of recording, so that you can alert the moderator that you're going to have to pause the recording in 5 minutes, or whatever the case may be.

I think that the least intrusive method of placing the mics is to use desk stands. Using mic stands and booms takes up too much room for four or more people, and

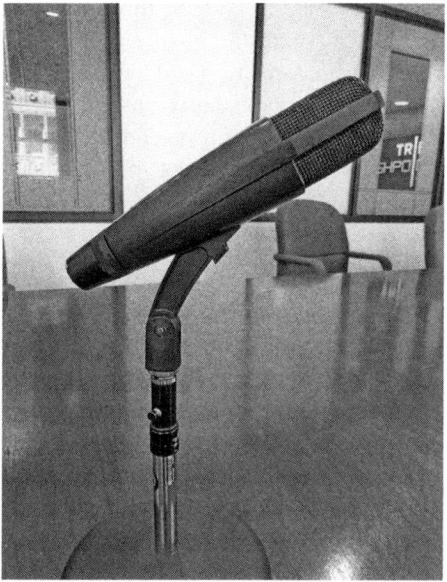

FIGURE 12.10

Atlas Desk Stand Setup

Source: Courtesy of Tribeca Flashpoint Media AA; photo by the author

they just get in the way; you want to make your amateur guests at ease, so avoid large stands if you can. Also, you should be aware that a large-diaphragm condenser mic can be quite heavy, so make sure that the desk stand that you select can support the weight of the mic without tipping over.

LIVE MIXING OF ROUNDTABLE DISCUSSIONS

One of the biggest challenges of recording roundtable discussions is the excessive amount of room sound that can be introduced into the recording as a result of having a number of open microphones recording at once. One mic will capture the noise floor of a room, and two mics will double the noise floor. If you have four or five (or more) mics open at the same time, you can easily see how the noise will be multiplied and lead to an unacceptable recording, not to mention the phase problems that will be introduced with the spacing of the mics. There are a couple of strategies that can be employed to help with this problem.

Doing this type of recording can be stressful and challenging, because you will be performing a live mix of the balance of the microphones to help combat the noise-floor issue. That is, when one person is speaking, the other microphones will be lowered in the mix, so that they don't add to the problem. However, by lowering the fader on the other mics, you stand the chance of missing the first word or two of someone who jumps into the conversation suddenly, and this will be very apparent in the final recording. You must be constantly on your toes and aware that any of the other participants may join in at any time. Fortunately, in this situation, most people will give visual cues that they are about to begin speaking: raising a finger, taking a deep breath, leaning forward in the chair, and so on. If you are constantly watching the participants closely, you can pick up on these visual cues and begin (slowly and smoothly) opening that person's mic. Then, of course, you would close the mic of the person who just finished speaking. One mic that I prefer to leave open most of the time would be the moderator's, because the moderator is most likely to jump in at any time, and we should always be cognizant of that.

One highly effective way to record these discussions is to record to multitrack and put each mic on a separate channel. In the editing stage, we can then mute those mics not being spoken into and thus avoid excess noise-floor issues. Care should be taken when performing this editing that we don't leave a gap (silence) between the edits; the lack of room tone becomes very noticeable, and all of your edits should be seamless to combat this situation. If you are multitracking the session, it's a good idea to place a different-color windscreen on each mic or to use colored mic cables, so that you can easily grab the fader for whichever participant is speaking, thus making your life much easier. The color-coding system is an invaluable technique

FIGURE 12.11

Multitracked Interview with Track Muting

that can carry over into many other recording situations and help you keep everything in the studio under control and running efficiently.

As you can see, interviews and roundtables have their own particular tricks and techniques. As with any recording situation, these will only come with experience. By mastering these skills, you put yourself in a position to be first call for any of this type of work and strengthen your professional portfolio.

More Voiceover Opportunities

13

Regardless if you are a recording engineer or a voice performer who records yourself in your project studio, the fact remains that we all want a chance to practice our craft and have a good amount of work. This can necessitate some out-of-the-box thinking about what type of work might be available in the world of capturing the human voice and recording voiceovers. Remember that the world is full of voice messages of all types, and keep in mind that someone, somewhere, must provide the material. Here is a short list of only some of the possibilities that you may encounter.

VOICE RESPONSE

When you call an insurance company, a customer support number, your doctor, or any number of other businesses and services, chances are you will be walked through a menu of options to help get you to the right person to talk with. Recording all of these voice messages can be a nice additional piece of business for you, and the opportunities are limitless if you can capture the work. We've all heard this type of recording before: "For software questions or problems, press 1," and so forth. These voice prompts are relied on by an increasing number of companies, both to reduce the number of phone operators they must employ and to help manage the customer's requests and get them to the proper person, so that the customer can get the information they need. For a number of years, I was involved with a company that provided health insurance information to a large number of insurance customers, and, as part of this service, we did voice prompts. The scripts were quite long and were constantly changing, and the voiceover artist developed a critical skill in performing these scripts. The lines of script were often of this type: "You have 10,719 dollars and 21 cents in your account." As the customer would add to and withdraw from the account, these figures would change. The trick in doing these voice response messages was to keep things sounding as natural as possible; we've all heard terrible mismatches when the figures are inserted, with the inflections all over the map. The amount of concentration required from the performer to keep everything flowing smoothly is quite remarkable to witness. For the above line, the recording is accomplished by first reading "You have" as a stand-alone line, followed with recording "left in your account." "Dollars" and "cents" are recorded separately, as well. Now, the tricky part: all of the numbers that are going to be inserted must be recorded individually, and this requires recording counting from one to ten thousand—not the most exciting of sessions, but one that kept us all on our toes. The performer had to ensure that "five hundred and" would match up with "seventeen" (at which point "dollars" would be inserted), to be followed with "and," then "no" and "cents." For an amount higher than ten thousand, the system was programmed to pull "twenty" or "thirty" or the appropriate number to insert into the sentence. As I said, this tested our concentration, and the work was quite tiring as a result. This is certainly not a time when the engineer, or the voice talent, can put everything on autopilot! When a break was called for, it was essential to play back some of the previously recorded lines, in order for the performer to find the rhythm and tone again. If the talent started to slip in the delivery and drift off from where she was 10 minutes before, it was my job to catch this and make the necessary corrections, often by backing up 10 or 15 minutes in the recording and playing that back for the voice artist.

With this type of work, as I mentioned, the most important criterion is concentration. You must be keenly aware of each word as it is spoken and how it relates to the other words that eventually will be joined together into a sentence for the listener. Consistency is the main watchword.

As for microphone choices, try to match the frequency range of the telephone, which by design has a narrow range. Either use a mic that has a good midrange emphasis or one that can easily be equalized into it. Avoid those microphones that are very round and have a pronounced low-frequency sound to them; they will only muddy the end result. I try to do a test recording, then patch that into the phone system and listen to the result over the telephone's earpiece. Alternatively, a laptop's speakers are not a bad representation of the sound the listener can expect to hear. I would consider doing some fairly heavy compression on my finished audio to keep my dynamics consistent and clear. Your work will be listened to by a wide variety of people, with a large range of hearing acuity, so plan for this at the beginning of the session.

PUBLIC ANNOUNCEMENTS

Under this heading, I've included any number of voice recordings that are played back to the public at large. If you think about all of the prerecorded announcements you are likely to hear in airports, train stations, supermarkets, and so on, you can begin to appreciate the opportunities for voice work in this general field. In the city I live in, the public transportation system utilizes announcements for each stop of the bus or commuter rail train as the stop is being approached, as well as announcements for delays and warnings against loitering, gambling, the playing of loud music, and so forth. If you keep your ears open as you move around the area where you live, you will become aware of all of these public announcements; with a bit of research, you can find what agency or other group is responsible for them and can make an offer to help record and master these announcements.

Once again, keep the final playback of your recordings in mind. Your audio will be competing with a wide range of noise elements in the environment, so it has to cut through all of the competition and be clearly heard and understood. As with telephone prompts, your audience will have a wide range of hearing, and you want to make sure that people of all ages and levels of hearing sensitivity are able to understand your message clearly. Once again, I would choose a microphone that is able to cut through the noise clutter, and also compress the dynamic range of the voice files fairly heavily to keep the dynamics under control.

WEB CONTENT

It seems as if every web site that we visit these days has some sort of audio content, and a number use voice to help get their message across. This could include, but is not limited to, commercials, animations, product training tutorials, customer or product announcements, and so on. The majority of automobile companies now make the owner's manual available as a DVD, and there a couple that are making the manuals available on their Internet sites. Someone has to narrate these, right? As with any other type of work, you have to do your research to find the right people to talk to in order to sell your voice or recording services, but this is an area that is already very large and growing daily. As part of your research, try to find as many web sites as possible that include audio in their content and identify how voice recording could benefit the company; then make your pitch. I predict that this area is only going to become more important in the future, so getting into the game and making the right connections now will be a big benefit.

Another type of web content that you can take part in is podcasting. These Internet-distributed shows can be entertainment-based, interviews, walking tours of your town, historical in nature, or any of countless other types of programming content. To get an idea of the wide range of podcasts already being done on a regular basis, go to the iTunes Store and search for "podcast." By listening to some of the content already out there, you can get an idea of the quality of the various shows, as well as content ideas. As an engineer, you can record, edit, and produce this content; as a voice artist, you can narrate as well. The one downside that I see with the podcasting world is the lack of a way to monetize your work—you have to have a very sizable, regular audience willing to subscribe to your channel in order to make a profit doing this programming, but it is fun and can offer an abundance of recording and editing experience. Many podcasts are being produced using a USB microphone directly connected to a laptop or desktop computer; although the result isn't top studio quality, most people are willing to accept the sound of the podcast, considering that the playback medium is often computer speakers, ear buds, or in a moving automobile, all of which are compromised as far as sound quality goes. Again, I would suggest listening to a wide range of available content to judge the quality issue for yourself.

One word about the quality of web-based audio: In the early days of the Internet, download speeds were painfully slow compared with today's high-speed access, and server sizes were small. Thus, the audio files delivered for web playback had to be quite small, and it wasn't unusual for the audio to be recorded at a sampling rate of 11 kHz, with 8-bit depth. Frequency and dynamic ranges were limited, and the audio was very poor. But, at least people could hear sound from the computer! Compared with the full-spectrum audio of 44.1 or 48 kHz, with 24-bit depth, employed today, it's hard to believe that audio on the web ever took off at all. With modern, high-speed Internet connections, we no longer have to compromise our product in the same way as earlier, and the days of 96-kHz audio are upon us. There is no reason to compromise audio quality during the recording process, even if most web sites use some sort of compression technology; now, we can make everything we do broadcast quality and to the highest standards.

VOICE TALENT DEMOS

One type of work that you can find, no matter where you might live, is offering a service recording demo materials for voice actors. You already have the microphones and the recording gear, and up-and-coming voice actors need to produce good-sounding demo reels to promote themselves. This is a nice, win–win situation: They end up with a top-quality product to distribute to producers and directors, while you pick up additional work and the opportunity to hone your voice-recording techniques. You'll probably want to locate a good source for background music and for sample scripts for the performers to read, so that you can end up with a finished-sounding product. Many communities have voice coaches or people who offer voiceover lessons, and contacting these folks to offer your services can pay off nicely. If you are a voiceover artist yourself, your knowledge can be passed on to the beginners in your field; if you are an engineer, use your experience with accomplished performers to help guide the newcomer doing the demo to a better performance. Either way, passing along our knowledge to others can be a source of real satisfaction in our work.

Today, most of the casting sessions for finding the perfect voice for a project are done online, and the voice talent competes on a global scale, thanks to the Internet. Yet even with this reality, those actors still have need to record, edit, and produce the demos that will be heard far and wide. For experienced actors who have done a large number of voiceover projects, their demo piece must still be compiled from examples of their previous work, and this is something that you might consider offering to them. Just as with a music album (if that word is still even used anymore), the various pieces must be "programmed" into just the right order, levels need to be adjusted and evened out, and the finished demo needs a mastering pass to make it into a coherent whole. Many voiceover artists are comfortable performing these tasks themselves, and, with digital editing software, there really isn't much mystery to the task, but it is often beneficial to have a disinterested third party (that would be you) to bounce ideas off. An actor editing himself doesn't have the necessary distance to gain a good perspective of the work.

AUDIO BOOKS

A large market segment of the recording industry lies in what is commonly known as "books on tape." Even though today the vast majority of this product is released on CD, DVD, or by digital delivery, the name is still with us as a reminder of the days when audiocassettes were the norm for listening to this material. In this segment, there are two broad categories of production: Either the author will read his or her work, or a voiceover professional is hired to voice the book. There are some instances of producing the book much like an audio play, with music, sound effects, and multiple characters, but this is rare owing to cost constraints.

The market for books on tape is very large and hasn't diminished over time; in fact, with digital delivery, sales are up in this market and can be a good source of additional revenue. Usually, the producer will be the publishing company of the original, written work, and the largest of these publishers will have their own in-house divisions set up specifically for overseeing the recorded versions of the books. The content can be anything from fiction to poetry, autobiography to history, and nearly anything else you can think of. Classics of literature have been recorded and released, as have the very latest thrillers and romance titles.

The books-on-tape market has some requirements in the recording that nearly all publishers demand: The studio should be very quiet and reflection free; a large-diaphragm condenser microphone is almost always specified; and the editing must be very, very precise. As you might imagine, just as with other types of recording mentioned in this chapter, a keen ability to concentrate is demanded of the engineer and the audio director, to ensure all of the words are pronounced correctly and nothing is skipped over. Here, as in so many other instances of recording the spoken word, proper note taking and documentation are of the utmost importance. It would be rare, indeed, for all of the recording to be concluded in one session; often, the completion of the recording is accomplished over a period of days, even weeks or more, and matching of tone and pacing has to be precise and not left to chance. Any unwanted sound, such as a page turn or an unusually large breath, must be noted for the editor to cut out, and any stomach rumblings or mouth sounds must be taken care of immediately by redoing the line or lines in question. One other thing to keep in mind is that the reader will probably find a comfortable rhythm to fall into during the read, and you don't want to interrupt that with too many starts and stops. If something must be redone for any reason, it may be preferable to make a note of it and then redo that section as a pickup, at a later time during the session (perhaps at the end of the chapter being recorded).

THE WACKY WORLD OF TOYS

You will discover the spoken word in all manner of unlikely places, and one of the places where you can have a good deal of fun is providing voice recordings for toys. Closely allied with recording for animation because of the use of character voices and sense of play, providing the voices of various toys can prove to be a good long-term project. Dolls, locomotives, fire trucks, board games, and more, often come with speaking parts built into them, and, if you can locate a maker of these products and a good character voice actor, this can be a nice addition to your recording résumé. Obviously, you need access to a manufacturer of the toy in question, but, if there are any in your area (or you can provide good audio files long distance, via the Internet or ISDN capabilities), they are well worth cultivating. A clean signal path and the ability to convert the audio into the file types specified by the manufacturer are a necessity, as is flexibility in your normal recording routine. As in recording for animation, some amount of time will be required for the actor to find the correct character voice to match the toy, and this should be factored into your scheduling.

AUDIO TOURS

There is a large market in the world of audio tours for voiceover recording and production. Walking tours, bus tours, driving tours, and more, all can be an opportunity to practice our craft and provide production services. Most museums have audio tours that can enhance the viewing of the exhibits, usually in the form of either rented headsets or telephone handset-like devices that the patrons hold to their ear. Most often these days, the devices connect wirelessly to a server where the audio is stored, and visitors can access the narration for any stop in the museum by entering a number on a keypad. Likewise, walking and driving tours access the audio from an external source (including the Internet or a cell-phone app), and the audio begins when the appropriate location on the tour is reached. These types of tour can be produced with or without music and sound effects, but the information that the patron hears is, of course, paramount. To this end, the engineer must recognize the fact that the listener will be hearing the material through headphones or a speaker pressed to the ear; therefore, a clean and well-recorded program must be provided. Also, take into account the fact that the location the audio is to be heard in may very well be noisy: Driving noise, street traffic and pedestrians, museum patrons, and the like, all contribute to the noise floor of the space and compete for the listeners' attention. Therefore, a clean recording is a must.

SOMETHING COMPLETELY DIFFERENT

One of the more unusual requests that I have had for voice recording came in the form of a telephone call one day. A gentleman from the county Board of Elections wanted to know if I could talk with him about the feasibility of incorporating voice into the new generation of voting machines that the county was about to purchase for an upcoming election. The idea was to provide services for the sight-impaired when they went to cast their vote: A voice prompt of the office and who the candidates were that were running for that seat. I met with the election representatives at their offices to discuss the matter. I was shown a prototype of the new voting machine, and I had come prepared with a selection of voice demos from both male and female announcers. What I had imagined would be a pretty straightforward request began to mushroom into a major piece of logistic planning and documentation. First, I had to meet with a representative from the manufacturer to determine the audio file type that the machines would accept (.wav, .aiff, MP3, or some other format) and then learn the maximum file size that the machine would play back. Of course, how the voice files would be imported into the machines also had to be discussed, as did my delivery path of the files to the client (the county) for uploading into every voting machine that the county would be using during the upcoming election. I did a series of sample recordings and worked with the manufacturer to optimize the audio quality and delivery path of the files.

These technical questions were just the beginning of the complications as it turned out. Not only would the name of every candidate running for every office in the election have to be recorded and cataloged, but, because it was a general election, all other offices (local, state, and national) would also have to be included. Because this particular county is one of the largest in the United States, encompassing 128 municipalities, each of which has multiple elected offices, you can now begin to imagine the scope of this project. The surprises didn't end there: In my county, ballots are printed in a variety of languages to accommodate the large foreign-speaking population that makes up our electorate. Besides English, ballots are printed in Spanish, Polish, and Chinese. Each candidate's name, party affiliation, and office would have to be read four times, once for each language.

Owing to my previous work on many voice-prompting projects, I helped to figure out the most efficient path for recording all of this material. If the line of copy was to read, "For County Dog Catcher, in the Independent Party, Thaddeus R. Futzenhauser," I knew that, no matter how many candidates were running for a given office, we would only have to record "For County Dog Catcher" one time, and the same for the party name, which would save quite a bit of time and money. It would also help to reduce the individual file sizes. Checking with the manufacturer of the voting machines, I determined that the machines could perform this linking function if properly programmed. Next, on top of figuring out how long all of this would take to complete, and how much it would all cost the county, I was now tasked with arranging for native speakers of these other languages to be auditioned and recorded. The script for this job resembled a large phone book!

I came to the conclusion that this was beyond the scope of my freelance engineering business, and that there was no way I could accomplish all of this by myself, and so I moved into the role of supervisor and spoke to a friend who is the audio director at a large production company with multiple recording studios. We worked out a plan for how we could meet the requirements of the contract and the time deadline for the upcoming election. Needless to say, there were multiple engineers working at once, with each language being handled in its own room, and with a couple of assistant engineers checking for content, pronunciation, and editing precision, and then naming and delivering the files. In the end, it all went off without a hitch, but with a lot of late nights and weekends. Yet another example of voice-recording opportunities coming at you from unexpected places.

IN CONCLUSION

All of these ideas and many more contribute to your potential base of voiceover work. Your local community will determine the amount and types of content that you can consider, but the important thing to remember is that there are many, many more opportunities for voiceover than just commercials and corporate narrations. Get out and explore what your community has to offer and be first in line to make a pitch for some of this business.

That's a Wrap

14

One microphone, one actor, and one channel on the mixer. What could possibly be simpler or more straightforward? That's the question that I started this book with, and by now we've seen that the act of recording a voiceover is anything but simple and straightforward. Recording sound is a technical field, and you have to know the technical aspects of your job—that's expected of you by the producers, directors, and actors. After all, that's why they hire you. But, as this book has shown, there's more to it than just the technical aspects. I think that, if you concentrate solely on the technical issues, in the end you are a slave to that technology. I prefer to think that the technology is here to serve you, and is nothing more than a useful tool that helps you get to the place where you want to be with your creativity. I believe that you should strive to reach that point where the technology disappears and you no longer think about it; then and only then can you give full rein to your creative ideas and let them flow naturally, without worrying about which button to push or which keyboard command performs the function you want. And, most people with whom you work will have no idea of what you do and can't speak the language to give you direction; they hire you because of your creative acceptance of their ideas and look to you to help make the project a success. You demonstrate your dedication to making this happen all through the process and you build a sense of trust between everyone involved; everyone involved works toward one goal—bringing the project to a successful completion. The majority of them don't care what technology you're using, or which version of software you are running; it doesn't matter at all to them, as long as you can bring their vision to a creative outcome, and a large part of that success comes from things other than your facility with your tools. These people are expecting your creative input and a solution to the problem at hand.

It's my hope that the information that I've presented in this book will help you with the skills needed to be a successful voiceover recording engineer. By following some of the advice found in this book, you will grow your recording career and expand your opportunities in the industry. Keep in mind that we all have our own preferred method of working and our own ideas on how best to arrive at an outcome that everyone involved sees as being the best that the project can be. What works for me might not be how you would approach things, and that's all right. If you only accept some of the ideas I've presented here and ignore others, that's fine— you still will have picked up some useful tips to make your voiceover recordings more effective. These differences are what make the world an interesting place to live in, and can be the source of endless conversation and thought. It is through our thinking about the various ways we can approach the task at hand and talking about it with others that learning takes place and our skill set evolves. Be open to learning from everyone—keep your brain turned on and listen for subtle hints from the people with whom you're working. Don't immediately discard ideas from others because they're not your ideas. Even though the person you're working with doesn't speak the technical language and verbalizes an idea in abstract terms, take the time to consider what she's saying. Don't be afraid to experiment, to try something off the wall, and to accept some mistakes as brilliant creative solutions. Sometimes, some of our best work comes about because of what we initially judge a mistake. If you analyze your work critically, you might see that, although you meant to do one thing, something else has happened, and this sometimes comes from a subconscious artistic or creative impulse that you were unaware of in your conscious thought. Once again, if you can get the technology out of the way of your thought processes, these subconscious ideas can flow out of your fingers and into the session, without your being aware of them.

BUILDING THE SENSE OF TRUST

Of course, the technical things matter a great deal; the more you know about acoustics and room design, the better your choices will be on microphone placement. The more you come to know about mic design, the better you will be able to choose just the right mic for a particular individual. The better your editing skills, the more the direction you give during a session will help you in the editing later. All of this is very important and is the basis for your entire career. Some engineers are more technical than others, but, for anyone, a good grounding in these topics is essential. As I just mentioned, this technical understanding is why you were hired in the first place and is your primary function in the studio; anything else that you can bring as an individual to the session is the reason the clients keep coming back to you. Working on these skills at the same time as you are trying to develop all of the other things we've talked about in this book can seem overwhelming, but it comes with experience and time.

You should be aware that the most important equipment that you have in doing the job at hand doesn't involve computers, or software, or editing platforms. We all possess this equipment, and I'll remind you to always have it at the ready throughout the session—*your ears*. Before you get all involved with figuring out everything else that has to be accomplished in the session, make very sure that you are critically listening to what is going on, both what's coming over the microphone and what others in the control room are saying.

A large part of your technical knowledge revolves around the subject of release formats. Of course, you have discussed what the release format is going to be before the session started, as part of your initial conversations with the client. You should be able to deliver whatever the client requests and know the difference between the various formats, so that you can help guide the client toward the best solution for his aims. Learn why there is a difference between 44.1-kHz and 96-kHz sampling rates, between 8-bit and 24-bit audio, and in what situation you would choose one over another. Can you quickly and efficiently change an audio file from a .wav format to an .aiff one? Why do we not recommend final delivery based on the MP3 format? What would you suggest instead? All of these questions are why the client comes to you and trusts you to give her the best solution for her needs, and this is all part of your job as an audio professional.

As I've mentioned often throughout this book, the main thing that you bring to the table is a sense of trust, and that trust must be earned every day. If someone asks you if you can get this project mixed and delivered by noon tomorrow, the answer is either yes or no. If you answer yes, be sure you can do it (even if you have to stay up all night to finish), and then make sure that you actually do it. If the honest answer is no, explain why and ask if there is any possibility of an extension. Either way, be up front and honest with your answer. If what they are asking can't be done, offer possible solutions—once again, this shows your dedication and concern for their project and demonstrates that you have their interests in mind; you are interested in helping them solve the problem. You are building that sense of trust. A few years back, I was approached about doing the audio post for a 60-minute documentary. In the initial conversation with the client, I discovered that he wanted it completed in five days! I could have told him he was out of his mind to expect that kind of turnaround and walked away; after all, working by myself, a film of that length would usually take three weeks, at a minimum, to complete the amount of work that it called for. I asked for an extension and was told that there was a firm deadline on this film, and that was all the time available to finish it. I was honest with the client and explained why I would only have time for a basic sound job, with no bells and whistles and very little creative sound design, but said that, yes, I could give him something resembling a finished film in that time frame. Well, in the end, it took me six extremely long days (the delivery was on a Monday, so that helped with the extra day), and the client was quite pleased with the result—as was I, as it turned out. Sure, there were many places that, to this day, I want to revisit and sweeten, but the film ended up winning a number of prestigious awards, and the client has come back with other projects, and some of them actually have reasonable time frames. Through my being honest about what could be accomplished and not leaving the impression that a more creative and fuller soundtrack could be accomplished, the client had trust in me to give him what I had described, and, in the end, everyone was satisfied.

As we have seen, a large part of the success of a project comes down to your project-management skills. The better you are at this, the better an audio practitioner you will be. By keeping on top of all the myriad details of the job—your documentation, microphone selection, archiving, and more—you will simplify your life, even if it might not seem like it at the time. After all, these various tasks take time, and it's easy to fall into the habit of skipping some of the steps in your project management. Don't get lazy! If you don't take the time to keep good notes on the take sheet during recording, your editing time will increase. If you fail to archive your materials properly, you will spend a huge amount of time finding the right pieces if you have to revise the project. Each of the steps is important and should be performed with due diligence. I don't mean to sound like I'm harping on this just because I might be obsessive–compulsive—I'm speaking from experience. Over the course of my career, I've missed something, only to have it come back and bite me later. The more attention to detail you bring to the job, the easier it becomes, and, before you know it, all of this attention to taking care of the small things becomes habit, and you don't have to think about them much, you just do them.

In this section, I've talked a bit about building a sense of trust with the people that you work with, and there's one other thing that I want to add. We all know that some strange things can happen during a session, and some extremely funny things as well. Be very, very careful about sharing some of these with others. *Never* post outtakes of a session without permission from those involved, no matter how funny you think they may be. Don't do anything that might be seen as embarrassing another person. Avoid water-cooler gossip about the people you work with, and don't bad-mouth other audio professionals to make yourself look better. If you do tell a story about someone, leave their name out of it—the story will have just as much impact, but you won't demean the parties involved. If you become known as a person who "tells stories out of school," no one will be able to trust that you won't soon be talking about them. And, *never, never, never* discuss one client's work with another client. At one point, I was doing commercial work for three competing banks, and everyone knew it, but I would never have dreamed of saying, "You should hear what Acme Bank is going to come out with next." How could my clients from the competing institutions trust that I wouldn't soon be talking about their upcoming campaigns? They trusted that I would keep my mouth shut about their business, and that's as it should be. Beyond trust, this also earns respect. At the end of the day, we only have one thing, and that is our good name—protect yours jealously. Once you violate trust, it is nearly impossible to regain.

One final point on this topic: We have to recognize that we're all human, and we all have days that are better or worse than others. There's nothing we can do about that, but what we can do is accept that everyone goes through this. What makes a true professional is the ability to work through the difficult times and still give it our best. If you sense when someone else may be struggling, for whatever reason, and then give them support, recognize them as being human, it helps to build this sense of trust. When others can tell that you are trying to help them and be supportive, the bond of trust is strengthened and gives everyone the confidence to do the best that they can.

YOUR INSURANCE POLICY

I have spent some time in this text discussing the importance of proper archiving of your work, as revisions of voiceover projects happen so frequently. You should get into the habit of archiving often—I suggest at least once a day for those projects that you've worked on during that day. If you do this and something were to happen to your hard drive, the most work that you would lose is one day's worth, and that's not the end of the world. The ultimate disaster is when the client calls to revise something, and you no longer have it! Trust me, it's happened.

Because we now live and work in a digital environment, there's another curve ball that just got thrown at us. In the old days of analog tape, once you had recorded and finished a project, all you had to do was file the master tapes away and you could retrieve them years later. Now, everything we do is software-based and software-dependent. This presents a real problem for the archive: If I were called upon to restore or revise an analog project, even if that particular format no longer existed and there were no machines left on the planet to play the tapes, it wouldn't be the hardest thing in the world to reverse engineer how a tape machine or a film projector works and come up with a way to play the old materials. However, now that everything is software-based, if you archive a project today, can you be certain that the necessary computers and software are going to exist a hundred years from now that would allow you to work with the materials? The answer, of course, is no. So, now we're faced with the task of transferring our materials from one format to another as new formats come along, to ensure that, at some point in the future, we can still access them if need be. We never can predict when we might need to work with them again. So, you have to carefully consider your archiving strategy to ensure that, in the future, all of these materials will be accessible.

One further word on the topic of archives and backup copies: As my friend Jeff is fond of saying, "If it only exists in one place, it doesn't exist." In fact, he goes so far as to say, "If it only exists in three places, it doesn't exist." Things have a habit of disappearing at just the wrong time, and, if you rely only on your hard drive to store your work, it's almost guaranteed that the hard drive will crash, and none of the data will be able to be recovered. This has happened at one point or another to almost everyone, and, if it hasn't yet, it will. Get in the habit of doing multiple backups. A little prevention can pay huge dividends. This book, for example, was written primarily on my laptop computer, but I had a continuous backup drive running while I was working on it, and, at the end of every writing session, I would connect a portable hard drive and copy my day's work over to that, and then finish off the writing session by also archiving to a dedicated thumb (or flash) drive. I put aside time in my day's schedule to complete these tasks and performed them diligently. Each of these portable devices was stored in a separate location, so that, if my office were to go up in flames, I had confidence that the book wouldn't perish. You should have the same level of backup protection for your audio work files. All it takes is one disaster to convince you of the importance of your backup copies. Don't be that guy who has to explain to the client that the project is no more. Storage is cheap; losing a client isn't.

To this point, I've talked about archiving your audio files, but, keep in mind that you also need to archive all of the relevant paperwork associated with each session; scripts, take sheets, delivery specs, telephone numbers of the key players, any video materials that you may have used, music licensing forms, lists of sound effects used, who you delivered to and on what date (and which delivery method you used), and many, many more items. As stated earlier, the goal is to be able to resurrect the entire session at any point in the future, without skipping a beat. It goes without saying that, the better you are at properly labeling and storing all of this material, the more trust the client is going to have in you.

We all work in physical spaces that have finite amounts of storage room, and, at some point, you will be faced with culling your archives of those jobs that were done some time ago and that, in all probability, won't need to be redone. You should have a return policy thought out and presented to the client in the initial contract for the work. At a number of facilities that I've worked at, the policy was that we would store these materials for a specified period of time (for about three years, generally), and they then would be returned to the client. If the client requested that we continue to hold on to them, we would then charge a yearly storage fee (usually nominal). Remember that the client paid for all of the materials, and they belong to the client—there is no way that you can simply dispose of them whenever you would like to. The storage is a service for the client, and the materials should be treated as belonging to someone other than your studio. Knowing which materials are due to be returned is one of the reasons that we spent some time talking about proper storage methods, with a database to track the items and a clear marking of the individual session date and tracking number on the materials themselves. Once again, attention to detail at the time of the session makes your life much easier in the end.

A PASSION FOR VOICE

I started this book by telling you about the time I was about seven or eight years old and the day that I discovered the magic in the human voice. To this day, I can vividly recall listening to that radio play of *A Christmas Carol* and how I felt as I discovered the storytelling ability of just the voice, unencumbered by pictures. I didn't realize it at the time, of course, but that day started me on a path to a life in the studio, recording and producing all manner of voice items, and it is out of these experiences in the studio that this book was created. As we have seen, there are many, many types of voiceover recording to be done, and each requires your thoughtful attention. By honing your skills and working on them in every session, you will begin to be seen as a person who cares about each individual project that is presented to you, and you'll begin to earn the trust and respect of the people you work with in this highly collaborative field. From that, you may form long-lasting personal relationships that enrich your life in many unexpected ways. The first day that I showed up for work, at the very first studio that took a chance on me as a wannabe engineer, I observed the owner of the studio recording a voice-over session. Both the producer and the voice talent from that very first session remain friends to this day, and, over a period of years, we spent untold hours of creative exploration together. Respect was gained on both sides, and we still retain that respect to this day. Both are by now long retired from the industry, but we still communicate and let each other know how we are doing and the paths that our lives have taken.

By developing your skills in voiceover recording, by understanding your clients' needs and projects, by coming to understand how actors work and are motivated, and by not treating a voice session as just something that has to be endured while waiting for your next dream session; by giving proper attention to what you are doing at all times and by not setting up a rock-and-roll vocal microphone and having the attitude that it's good enough—you will become known as *the* expert on voice recording in your market and will see your business and personal relationships grow and prosper. Be that person—understand the importance of human-to-human communication and the humanity in everyone you work with, and your world will expand beyond measure.

Once you come to realize the subtleties and the art of voice recording, you can become the person—the *only* person—that your clients and actors want to work with. You are the one who seriously gives thoughtful and creative attention to what they are trying to accomplish.

In any field of endeavor, the one trait that all successful people have in common is a *passion* for what they do. That passion is your key to success. Develop it, nurture it, and come to understand that communicating with other human beings is the highest form of our existence. To this end, know that there is no small or insignificant session; every day, every session you do is the most important thing you've ever done. I'll leave you with some wisdom from a very talented and caring man, a studio owner who worked in the film sound business, by the name of Ric Coken. Unfortunately, Ric passed away just a few years ago, but I will never forget what he once told me: "When people ask what is the best session I ever worked on, my answer is always the same—the one I'm working on right now."

Resources

The following list of books and web sites may prove useful to those looking for more information on the subject of voiceover recording as covered in this text. Although this listing contains many sources of this information, there are many more to be discovered with your own research, and I would appreciate hearing from you on others that you find relevant and helpful, so that I might share them with others.

BIBLIOGRAPHY

These books were among those that I found particularly helpful while doing the research to put this text together:

Alten, Stanley R., *Audio in Media*. Boston, MA: Wadsworth Cengage Learning, 2011.

Benade, Arthur H., *Fundamentals of Musical Acoustics*, 2nd, revised edn. New York: Dover Publications, 1990.

Davis, Gary and Jones, Ralph, *Sound Reinforcement Handbook*. Milwaukee, WI: Hal Leonard, 1989.

Everest, F. Alton and Pohlman, Ken C., *Master Handbook of Acoustics*, 5th edn. New York: McGraw-Hill, 2009.

Gallagher, Mitch and Mandell, Jim, *The Studio Business Book*, 3rd edn. Boston, MA: Cengage Learning, 2006.

Gertner, Jon, *The Idea Factory: Bell Labs and the Great Age of American Innovation*. New York: Penguin Press, 2012.

Hampton, Dave, *The Business of Audio Engineering*. New York: Hal Leonard Books, 2008.

Hogan, Harlan and Fisher, Jeffrey P., *Voice Actor's Guide to Recording at Home and on the Road*. Boston, MA, Cengage Learning, 2009.

Holman, Tomlinson, *Sound for Film and Television*, 3rd edn. New York: Focal Press, 2010.

Huber, David M. and Runstein, Robert E., *Modern Recording Techniques*, 8th edn. New York: Focal Press, 2014.

Nisbett, Alec, *The Sound Studio*, 7th Edition. Burlington, MA: Focal Press, 2003.

Wright, Jean Ann and Lallo, M.J., *Voice-Over for Animation*. Boston, MA: Focal Press, 2009.

Yewdall, David Lewis, *Practical Art of Motion Picture Sound*, 4th edn. Waltham, MA: Focal Press, 2012.

USEFUL WEB SITES

These web sites may prove to be useful for you in finding more information on the subject of voiceover recording. Note that I am not including addresses for most individual manufacturers: You can search Google.com to find their home pages. Also, be aware that this is only a suggested listing of web sites: There are many others waiting for you to discover, depending on your individual needs. I have organized these sites according to subject for easy reference:

Acoustic Materials

*   Acoustic Sciences Corporation—www.acousticsciences.com
*   Auralex Acoustics—www.auralex.com
*   Sonex Acoustical Products—www.sonex.com

Audiobook Publishers

*   Book Business—www.bookbusinessmag.com
*   Audio Publishers Association—www.audiopub.org
    This address is the home site of the Audio Publishers Association and is a very useful place to find information on this subject, as well as links to other writing on the audiobooks business. Other addresses for audiobook publishers can be found by searching for individual publisher's web sites. Note that not all publishers of printed material create audiobooks, but, by searching out individual publishers, you will be able to find many that do. Also, you can search Amazon.com or Google.com for an idea of which publishers are involved with audiobooks.

Building a Booth

Here are a few articles that provide useful information on the construction of a voice-recording booth:

*   http://scottreyns.com/building-a-vocal-booth.php
*   www.backstage.com/advice-for-actors/voiceover/how-build-your-own-recording-booth
*   www.primacoustic.com/app-vocal-booth.htm
*   www.soundonsound.com/sos/mar06/articles/studiosos.htm
*   http://thevoiceoverguide.com/tools-necessary-for-working-professional-online-voice-talent/chapter-1–5-sound-booth
*   http://voiceoveraudio.blogspot.com/2011/07/voice-over-booth-build.html
*   www.peterdrewvo.com/html/setting_up_an_affordable_home_.html

Copy Stands and Lighting

- Manhassett Specialty Co.—www.manhasset-specialty.com

Ear Training Sites

- www.8notes.com/games
- http://c-allen.co.uk/cart/index.php?route=product/category&path=60
- http://harmanhowtolisten.blogspot.com

Prefab Booth Manufacturers

If you would rather purchase a prefabricated recording space, these manufacturers give you all of the necassary information on their web sites:

- Porta-Booth Pro—http://voiceoveressentials.com
- StudioBox GmbH—www.acousticbooth-studiobox.com
- VocalBooth, Inc.—www.vocalbooth.com
- WhisperRoom, Inc.—www.whisperroom.com

Recording Equipment

- B&H Photo/Video/Pro Audio—www.bhphotovideo.com
- Full Compass Systems, Ltd.—www.fullcompass.com
- Guitar Center Pro—www.guitarcenter.com/GC-Pro-Home-g27135t0.gc
- Sweetwater Sound, Inc.—www.sweetwater.com
- Voiceover Essentials, Inc.—www.voiceoveressentials.com

Telephone Interfaces

If you are looking for information on manufacturers who provide telephone interfaces for remote sessions, check here:

- Audio TX—www.audiotx.com
- Source Elements, LLC—http://source-elements.com/source-connect
- Telos Alliance—www.telos-systems.com

Further Exploration

This listing of web sites may be of use to you in gathering further information on the techniques of recording and performing voiceover:

- Designing Sound—www.designingsound.org
  This site has a large amount of information on the subject of sound design and is updated as often as four times a week. Although much of the information on the site does not pertain to voiceover directly, there are many articles and postings

on the subject, including an ongoing series of articles on game audio and voice recording. I highly recommend becoming familiar with this site.

- SAG/AFTRA—www.sagaftra.org
  The homepage of SAG/AFTRA and very useful for information on the main voiceover union. From this page you can search for local chapters in your area.
- Voices.com—www.voices.com
  This is a resource that has thousands of voice talent demos and is also a clearinghouse site that has job listings available for voice actors. This is one of many online casting sites that you can list your demo on. As with any enterprise, you will find that some sites are more productive and useful than others, so let the buyer beware. Also, search for talent agencies in your area.
- Voiceover Info—www.voiceoverinfo.com
  A producer of voice demos for actors looking to break into the voiceover business. The homepage of Sound Advice, Inc., here you will find information and coaching services available either in person or via Skype.
- Voiceover Times—www.voiceovertimes.com
  A useful listing of news updates on the voiceover business, aimed at the working actor.

These are only a very small sample of web sites that you can find with a browser search. Depending on your area of expertise (engineer, video game actor, character actor, and so on), a more specific search will yield a large number of places for you to explore in the world of the voiceover.

Index